Andrew Wilson

Sketches of Animal Life and Habits

Andrew Wilson

Sketches of Animal Life and Habits

ISBN/EAN: 9783337014124

Printed in Europe, USA, Canada, Australia, Japan

Cover: Foto ©berggeist007 / pixelio.de

More available books at **www.hansebooks.com**

SKETCHES

OF

ANIMAL LIFE AND HABITS

BY

ANDREW WILSON, Ph.D., &c.

> Oh, what an endlesse work have I in hand
> To count the sea's abundant progeny!
> — Spenser.

W. & R. CHAMBERS
LONDON AND EDINBURGH
1877

Edinburgh:
Printed by W. and R. Chambers.

PREFACE.

THE present series of Sketches has been compiled with the view of affording general readers, and especially the young, some popular and yet trustworthy ideas regarding some of the most interesting groups of the animal world. The work may in fact be regarded as a natural history text-book adapted for use in Nature's school at large, and as a guide to the use of the observant powers, through the due exercise of which all true ideas of Nature are acquired. Some of the Sketches have already appeared in *Chambers's Journal*, and are reprinted here with some additions, and with such alterations as were found to be necessary in adapting them for volume-form. The papers lay no claim to originality; but the day of grave errors in popular scientific works is not yet past, and the author would fain hope that the Sketches may at least be found to possess the merit of correctness. The highest purpose this little work may serve is, that of fostering a love for the personal and direct observation of Nature. Knowledge so acquired invariably brings its own reward in the power it confers of discerning higher aspects in 'all the workes of Nature.' And as a result of such studies, we may say with Wordsworth of the most commonplace object around us, that we greet it

> Like a pleasant thought,
> When such are wanted.

CONTENTS.

	PAGE
A PEEP AT ANIMALCULES	7
LIFE IN THE DEPTHS	25
CONCERNING SEA-ANEMONES	39
SEA-EGGS	53
A GOSSIP ABOUT CRABS	63
SHELLS AND THEIR INMATES	72
BUTTERFLIES OF THE SEA	87
CUTTLE-FISH LORE	93
ON SOME ODD FISHES AND THEIR COMMONPLACE NEIGHBOURS	113
SOME CURIOSITIES OF INSECT-LIFE	137
ON SOME CURIOUS ANIMAL COMPANIONSHIPS	149
ANIMAL DISGUISES AND TRANSFORMATIONS	159
ANIMAL ARMOURIES	177
'FOOT-PRINTS ON THE SANDS OF TIME'	187

SKETCHES

OF

ANIMAL LIFE AND HABITS.

A PEEP AT ANIMALCULES.

IN the old days of natural history, the name 'animalcule' was applied in a very wide and general manner to denote all beings of small or minute size, irrespective of their structure or relations to other animal forms. The 'animalcules' of the older writers included a goodly number of minute animals of very varied and diverse kinds, and only when the microscope began to be perfected and the exact structure of animals carefully investigated, were the 'animalcules' arranged in an orderly manner. The old group containing these minute beings was in fact disbanded, and its members rearranged in divisions of the animal world often widely separated from each other by real and veritable distinctions. Thus, from being regarded as a name common to all minute organisms, the term 'animalcule' has fallen into abeyance in modern zoology; and is never employed in the present day without having a very distinctive prefix attached to it. We thus talk of 'Bell-animalcules,' of 'Wheel-animalcules,' of 'Infusorian animalcules,' and the like; so that while the name is expressive as heretofore of minute size, it carries with it a distinct meaning such as under the old system of naming animals it did not possess.

The 'animalcules'—that is, small or minute animals—belong to various groups of the animal world. Some are of low organisation, whilst others are of tolerably high structure. No study can afford a better or truer idea of the literal greatness of little things, than the investigation of some of the minute forms of animal life. We obtain in such a study glimpses of veritable worlds in miniature, each peopled by inhabitants of definite kind. Each leaf becomes a universe, each drop of water an ocean; whilst no less clearly does our study shew us that the dust of our pavements, the rain-drops in our gutters, and 'the green mantle of the stagnant pool,' when regarded by the light of nature-science, alike become vast repositories of life, every detail of which testifies eloquently to the wisdom of that Power which is truly *maximus in minimis*.

A phial of water, taken, along with a little of the green scum, representing the growth of lower plant-life, from the surface of a stagnant pool, will furnish the microscopist with abundance of material for many observations. And it may be regarded as not the least of the triumphs of science, in that, within the compass of a phial of stagnant water, it is enabled to descry wonders and mysteries which baffle the wisest, and puzzle the most profound in their attempts to explain many of the acts of the living beings with which such a world in miniature teems.

A description of a few of the objects which may be observed and studied by any one provided with a microscope of ordinary powers, and who is armed with a little previous knowledge and a modicum of patience and perseverance, may therefore prove interesting, in view of the profitable employment of a few hours of leisure time. And we may commence our study with the description of a very common, but at the same time most interesting and beautiful group of animalcules.

We may thus look upon a prospect which is well calculated to excite our wonder and interest. The eye sees a variety of form and structure presenting a combination of grace and delicacy hardly to be matched in the whole of Nature's domain. Within the compass of a small round disc or circle, we behold numerous beings, each consisting of a bell-shaped head, mounted on a delicate flexible stalk. The margins of the bells are fringed with minute processes, resembling miniature eyelashes, and hence named *cilia;* and these processes wave to and fro with an incessant motion, by means of which particles of solid matter suspended in the water around are swept into the mouths of the bells. Suddenly

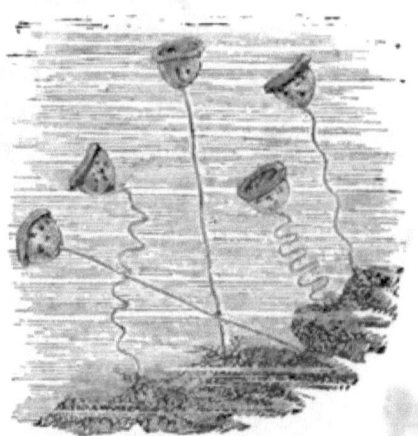

Fig. 1.—Group of 'Bell-animalcules.'

some impulse moves the beings we are gazing upon to contract themselves; and as if by magic, and more quickly than the eye can follow them, the bell-shaped bodies shrink up almost into nothingness by the contractile power of their stalks. Soon, however, as the alarm disappears, the beings once more uncoil themselves; the stalks assume their wonted and straight appearance; the little cilia or filaments once again resume their waving movements; and the current of life rolls onwards as before.

The spectacle we have been describing is not by any means a rare or uncommon one, to the microscopist at least. We have merely been examining, by aid of the microscope, a tiny fragment of pond-weed and its inhabitants, floating in a thin film of stagnant water. Attached to the weed is a colony of those peculiar animalcules known popularly as 'bell-animalcules,' and to the naturalist as *Vorticellæ*. These animalcules are readily procured for examination. Whole colonies of these and of neighbour-animalcules may be detected by the naked eye existing on the surface of pond-weeds as a delicate white nap, looking like some lower vegetable growth. And when a portion of the weed is placed under the object-glass of the microscope, numerous animalcules are to be seen waving backwards and forwards in all their vital activity. The general appearance of each animalcule has already been described. The bell-shaped structure which, with its mouth turned uppermost, exists at the top of each stem or stalk, is the body. The stalk is never branched in these particular animalcules; and except in certain instances to be presently noted, each stalk bears a single head only. The structure of the stalk is worthy of special mention. The higher powers of the microscope shew us that within the soft substance or *protoplasm*, of which not only the stalk but the body also is composed, a delicate contractile fibre is contained. This fibre possesses the power of contracting under stimulation, just as the muscles of higher animals contract or shorten themselves. And by means of this structure, therefore, the bell-animalcules, when alarmed or irritated in any way, are enabled to contract themselves with great rapidity, the stalk itself shrinking up into a spiral form. The whole operation, indeed, reminds one forcibly of some sensitive plant shrinking when rudely touched. The lower extremity of the stalk forms a kind of 'root,' by means of which the animalcules attach themselves to fixed objects, such as pond-weeds and the like.

The bell-shaped body is sometimes named the *calyx*, from its resemblance to the structure of that name in flowers. The edge of the bell possesses a very prominent rim, and within this we find the fringe of filaments or cilia, which in reality form a spiral line leading to the edge of the bell, in which at one point is situated the mouth, represented by an aperture or break in the rim of the body. We have seen that the cilia create miniature maëlstroms or whirlpools in the surrounding water, which have the effect of drawing particles of food towards the mouth. The study of the bell-animalcules affords an excellent example of the gaps which still remain to be filled up in our knowledge of the structure of even the lowest and commonest forms of life. No structures, for example, are more frequently met with in the animal world than the delicate vibratile filaments or *cilia*, so well seen in the bell-animalcules. The microscopist finds them in almost every group of animals he can examine. They are seen alike in the gills of the mussel and in the windpipe of man, and wherever currents of air or fluid require to be maintained and produced. Yet when the physiologist is asked to explain how and why it is that little microscopic filaments—each not exceeding in many cases the $\frac{1}{5000}$th of an inch in length, and destitute of all visible structure—are enabled to carry on incessant and independent movements, his answer is, that science is unable, at the present time, to give any distinct reply to the query. No trace of muscles is found in these filaments, and their movements are alike independent of the will and nervous system of their possessor; whilst, when removed uninjured from the body of the animal of which they form part, their movements may continue for days and weeks together.

What a field for future inquiry may thus be shewn to exist, even within the compass of a bell-animalcule's history —these animalcules being themselves of minute size, and even when massed together in colonies, being barely perceptible to the unassisted sight!

A very simple and ingenious plan of demonstrating the uses of the cilia in sweeping food-particles into the mouths of the animalcules, was devised by Ehrenberg, the great German naturalist. This plan consists in strewing in the water in which the animalcules exist, some fragments of coloured matter, such as indigo or carmine, in a very fine state of division. These coloured particles can readily be traced in their movements, and accordingly we see them tossed and whirled about by the ciliary currents, and finally swept into the mouths of the animalcules, which appear always to be on the outlook, if one may so term it, for nutritive matters. Sometimes when we may be unable to see the cilia themselves on account of their delicate structure, we may assure ourselves of their presence by noting the currents they create.

The structure of the bell-animalcules is of very simple and primitive kind. The body consists of a mass of soft protoplasm—as the living substance of the lower animals and plants is named; but this matter is capable of itself of constituting a distinct and complete animal form, and of making up for its want of structure by a literally amazing fertility of functions. Thus it can digest food; for in the bell-animalcules and their neighbours, the food-particles swept into the mouth are dissolved amid the soft matter of the body in which they are imbedded. And although the animalcules possess no stomach or any other internal organs, the protoplasm of the body serves them in lieu of that apparently necessary apparatus, and prepares and elaborates the food for nourishing the body.

We have also seen that the animalcules contract when irritated or alarmed. A tap on the slide of glass on which they are placed for microscopic examination, initiates a literal reign of terror in the miniature state; for each animalcule shrinks up as if literally alarmed at the unwonted innovation in its existence. This proceeding suggests forcibly to us that they are sensitive—if not in the sense in which higher animals exhibit sensation, at least in much the same degree and fashion as that in which many plants are regarded as being sensitive. And where sensation exists, analogy would lead us to believe that some form of apparatus resembling or corresponding to nerves exercising the function of feeling, must be developed in the animalcules. Yet the closest scrutiny of the bell-animalcules, as well as of many much higher animals, fails to detect any traces of a nervous system. Hence naturalists fall back upon the supposition that the curious protoplasm or body-substance of these and other lower animals and plants, possesses the power of receiving and conveying impressions, just as in the absence of a stomach it can digest food.

The last feature in the organisation and history of the bell-animalcules which may be alluded to in the present instance, is that of their development. If we watch the entire life-history of these animalcules, we may observe the bell-shaped heads of various members of the colony to become broadened, and to increase disproportionately in size. Soon a groove or division appears in each enlarged head; and as time passes, the head appears to divide into two parts or halves, which for a time are borne on the single stalk. This state of matters, however, does not long continue; and shortly one of the halves breaks away from the stalk, leaving the other half to represent the head of the animalcule. The wandering half or head is now seen to be provided at either end with cilia, and swims freely throughout the surrounding water. After a time, however, it settles down, develops a stalk from what was originally its mouth-extremity; whilst the opposite or lower extremity with its fringe of cilia comes to represent the mouth of the new animalcule. We thus note that new bell-animalcules may be produced by the division of an original body into two halves, whilst they also increase by a process of *budding*. New buds grow out from the body near the attachment of the stalk; these buds in due time appearing as young Vorticellæ, which detach themselves from their parent, and seek a lodgment of their own.

The 'bell-animalcules' present good examples of some of the lowest forms of animal life. They belong to a great class named *Infusorian* animalcules, from the fact that these beings occur in 'infusions' of animal and plant matter, and especially wherever such matter is freely exposed to the atmosphere. The stagnant pool is nothing more nor less than an 'infusion' of leaves and plants on a large scale, and all forms of decaying or putrefying matter serve as a soil in which the Infusorian animalcules flourish and grow. To the question, 'Where do the animalcules come from?' science has a plain reply to give. The atmosphere contains large quantities of the 'germs' and adult forms of these beings, existing in a dried state, but only requiring to fall into some suitable place, there to develop into active Infusorians. The infusion of decaying matter is thus to the germs what a suitable soil is to the seed; and little is it to be wondered at that life is so plentifully diffused, when we consider the manner and method of its distribution.

The further examination of the fragment of pond-weed on which the bell-animalcules reside may probably result in the discovery of some other and nearly allied species of Infusorians. We are almost certain to meet with a very pretty form named *Carchesium*, which closely resembles its bell-shaped neighbours, save in that the stalks are branched; whilst *Epistylis* is another graceful being, differing, however, from the bell-animalcules, in that the stalks are branched, but are not contractile.

Close by, adhering to the weed we note an animalcule, shaped exactly like a trumpet, and whose margins are fringed with cilia which wave to and fro with great activity. It possesses a green body, coloured, it may be noted, with *chlorophyll*, the identical substance which imparts a green hue to plants; so that the animalcule before us, and the cells of the weed to which it is attached, curiously enough, derive their colour from one and the same substance. The 'trumpet-animalcule' is the '*Stentor*' of the naturalist; and it must be admitted that as far as the shape and form of its body are concerned, it is worthily named after the celebrated person of mythological fame, whose talents in the way of vocalisation were of such a marked kind. Presently we may note the 'trumpet-animalcule' to detach itself from the leaf, and we then observe that the cilia which, in the bell-animalcules and in the fixed Stentor, serve by the currents they create to draw food-particles toward the mouth, now act as swimming-organs, and propel the animalcule by their vibrations quickly through the water.

Rushing hither and thither through the miniature sea under the microscope, we may discern many different kinds of these animalcules. Here, for example, is the most familiar of them all, the *Paramœcium*, or 'Slipper-animalcule' as it is popularly named, from the possession of a somewhat shoe-shaped body. The margins of this body are fringed with cilia, by means of which it paddles its way along amid the crowd of jostling neighbours; and when it rests for a moment from its wanderings, we can see multitudes of the little solid particles which float about in the water, swept into the mouth of the animalcule by the ciliary currents.

At one extremity of the body of the animalcule we can discern a clear round space. Fixing our eye intently upon this space we see it apparently disappear, whilst after an interval it reappears; and the movements of contraction and expansion seen in this space continue uninterruptedly during the life of the animalcule. This pulsating space is the 'contractile vesicle' (fig. 2, A, c, c) of the Infusorians. Its exact nature is still a matter of doubt, but in attributing to this space the function of circulating some fluid by its contractions through the simple body, naturalists have probably made a near guess at the truth. And this opinion receives support from the fact that in many of these animalcules little canals or outlets have been seen passing from the contracting space outwards into the body-substance.

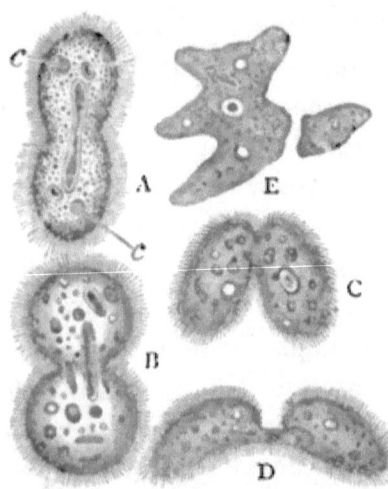

Fig. 2—Division of Infusoria: A, a paramœcium beginning to divide transversely; c, c, the 'contractile vesicles;' B, the same further advanced; C, vertical division of infusorian; and D, another mode of fission; E, an amœba undergoing division.

Amongst the crowd of slipper-animalcules which lie beneath our gaze, we can detect several exhibiting an apparently double shape of body, as if two small animalcules had united at once their interests and bodies. A short study of these apparently double individuals reveals an opposite state of matters; since, so far from there being any intention of effecting a union of interests, a process of separation and divorce is proceeding. Soon the double body resolves itself into two distinct halves, each of which goes on its way rejoicing, to pursue life and life's affairs on its own account. We have in fact been witnessing once again the process of *fission*, or multiplication by division, to use a somewhat paradoxical-sounding phrase. And if we scan the countenances of the animalcules around, we may see this process represented in all its stages. In some the body is just beginning to exhibit a slight notch (fig. 2, A); in others the separation has proceeded to a greater extent, B; whilst in others again, the partnership is on the point of dissolution, C; and the interested parties are about to bid a long and final farewell to each other. This process of the production of new individuals by the division of a single form is not the least wonderful feature of these animalcules, although it may be noted that in animals of very much higher rank —such as the sea-anemones and corals—a similar manner of reproduction prevails.

Making its way slowly along amidst the fragments of weed which impede its course, we can descry a peculiar

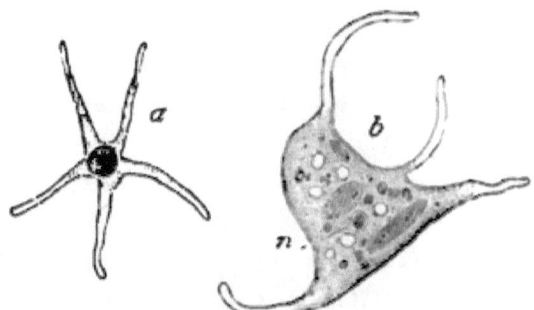

Fig. 3.—The Amœba or 'Proteus-animalcule:'
a, young *Amœba*, with five pseudopodia protruded; *b*, another and older specimen; *n*, the nucleus or central particle of the body.

shapeless and colourless being, which, as it moves, seems literally to flow from one shape into another. Movement in this case really means a constant alteration of the shape of the soft, clear, jelly-like body, which may not exceed the $\frac{1}{500}$th part of an inch in length. This is an example of the 'Proteus-animalcule'—the naturalist again drawing upon the lists of mythological heroes for a name, which in this case is expressive enough, as indicating the constant alternation of shape and form which the flexible Proteus was believed to evince. More frequently, however, the Proteus-animalcule is named the *Amœba*, a name derived from the Greek for 'change;' and under this newer title we may therefore recognise the 'Proteus-animalcule' of former days.

We note at once that this new candidate for notice differs from the bell-animalcules, the slipper-animalcules, and their neighbours, in its power of pushing out the soft protoplasm or jelly-like matter of its body into finger-like processes. When it moves, it thus pushes out a process of its body in the direction in which it means to travel, and the rest of the body flows, as it were, into this extended portion. And thus by continuous extension and contraction of its body it slowly moves amongst its fellows—a plastic, living speck, which ever and anon seems to fade away into nothingness, and to reappear before the observer's eye. When a particle of food consisting, it may be, of a minute plant, or of some other denizen of its native waters, comes in contact with the amœba, the latter launches out its soft body and literally ingulfs the morsel. The amœba exhibits, in short, the most convenient and primitive manner of taking food; since it receives food by any portion of its body, and ejects indigestible matters similarly through any part of its substance. It is a simple speck of living jelly, possessing no internal structures or organs, and yet living as perfectly, to itself and for itself, as the highest of living beings.

Occasionally, too, the amœba may exhibit the peculiar manner of producing its like, observed in the 'slipper-animalcule.' More than once I have watched an amœba stumble, as it were, across a fragment of pond-weed which lay in its path. The soft body was next extended both above and below the obstruction, apparently in the hope of surmounting the difficulty, but without success. The animalcule was, however, equal to the occasion; for the portion of the body above the obstacle parted company with that below, and each half moved triumphantly away from the spot, having converted a defeat into a veritable crown of success, in that two individuals were thus produced by the temporary difficulty of one.

A further search amongst the contents of our phial may result in the discovery of some animalcules of different

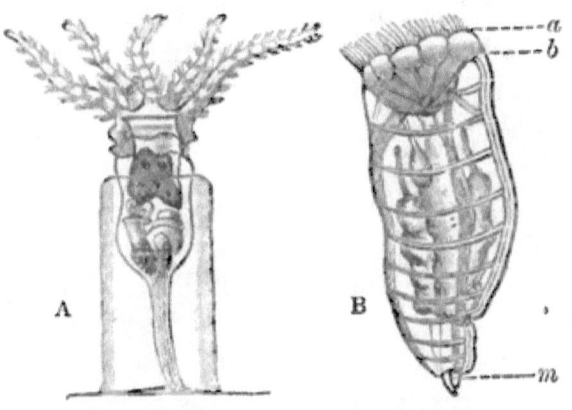

Fig. 4.—Wheel-animalcules or Rotifera:
A, *Stephanoceros*, a fixed rotifer; B, *Hydatina*, a free swimming species;
a, *b*, the ciliary discs; *m*, the terminal 'nippers.'

structure and of higher grade than the amœba and its neighbours. In the year 1702, Leeuwenhoek, a celebrated Dutch naturalist, noticed that some rain-water which had collected in a leaden gutter situated at the front part of his house exhibited a red colour, and with the commendable curiosity of true science resolved to investigate the cause of the unusual appearance. 'I took,' he tells us, 'a drop of this water which I placed before the microscope, and in it I discovered a great number of animalcules. Some of them red, and others of them green. The largest of these viewed through the microscope did not appear bigger than a grain of sand to the naked eye, the size of the others was gradually less and less: they were for the most part of a round shape; and in the green ones the middle part of their bodies was of a yellowish colour. Their bodies seemed composed,' he continues, 'of particles of an oval shape; they were also provided with certain short and slender organs, or limbs, which were protruded a little way out of their bodies, by means of which they caused a kind of circular motion and current in the water: when they were at rest, and fixed themselves to the glass, they had the shape of a pear with a short stalk. Upon more carefully examining this stalk, or rather this tail, I found that the extremity was divided into two parts, and by the help of these tails, the animalcules fixed themselves to the glass: the lesser of these appeared to me to be the offspring of the larger ones.'

In such simple language did this worthy old naturalist announce the discovery of an entirely new group of animalcules. In 1675, it is also interesting to observe, this same Leeuwenhoek had described the first 'bell-animalcule;' whilst his good fortune of 1702 has handed down his name to the present day as the discoverer of the 'wheel-animalcules,' or *Rotifera*—the species which had coloured the rain-water in the gutter, being the common *Rotifer vulgaris* of modern zoologists.

Our stagnant pools teem with these animalcules, and in many points their study assumes a deeper interest than that of the Infusorians. When we see a common 'wheel-animalcule,' our attention is at once arrested by the curious spectacle of the animalcule's possessing a pair of revolving wheel-like organs (fig. 4, B, *a, b*) at its head-extremity. We notice that the wheel-like structures constitute its organs of motion, and that by aid of the apparently revolving *cilia*

—which are thus again brought under our notice—the animalcule paddles its way through the yielding waters. When alarmed, it retracts or withdraws its wheels, but expands them in all their activity when all cause for alarm may be presumed to have passed away. The tip of the body or tail-extremity is provided with a little forked appendage (fig. 4, B, m), the use of which is to moor or anchor the animalcule to fixed objects. In the latter case the active wheels, which had previously propelled the organism through the water, serve as food-providers, since they create currents in the surrounding water, and these currents come laden with food-particles which are quickly swallowed and duly digested.

A closer study of the wheel-animalcules, however, would reveal to us the notable fact, that their revolving discs are wheel-like only in name. The appearance of rotating wheels, which forms so remarkable a feature in their history, is produced simply by the cilia with which the organs are fringed bending each in its turn, and with unerring regularity. The organs do not revolve, and there is, in fact, no rotation whatever, represented in the case. The illusion presented to us in the present instance, resembles that seen on looking at a field of golden grain in autumn, when the wind sweeps across the surface, and by causing each stalk to bend in its turn, gives us the idea that definite waves of movement are passing over the field. The 'wheels' of the animalcules are as fixed and as stable as are the rooted stalks of grain; the illusion so plainly discovered in the case of the corn-field, being as susceptible of explanation in the case of the wheel-animalcules.

The largest of the Rotifers measures about the $\frac{1}{36}$th of an inch in length, and many of these animalcules appear as veritable giants when compared with their Infusorian neighbours. But it is noteworthy to observe, that with superiority in size has been developed a marked advance in structure. Within the body of the 'wheel-animalcules' we find not only a distinct stomach and digestive apparatus (fig. 4), but also a set of active jaws for breaking down the food-particles, a system of water-vessels, definite sets of muscles, and a nerve mass, of very large size when compared with the bulk of the body. The delicacy of their bodies may be better imagined than suggested or described, and in view of what still remains to be told regarding the wheel-animalcules, the possession of complicated organs only renders their biography the more curious and surprising.

The astonishing fact has long been known that the wheel-animalcules, active and highly organised as they are, may be dried artificially, or naturally by the heat of the summer sun, from the water of the pools in which they live, and that in this mummified state they may be blown far and wide, as mere dust-specks by the winds. We may once again ask Leeuwenhoek to relate his experience of the marvellous powers and qualities of his foster-children. In a paper on the Rotifera, contributed to the Royal Society of London, he says: 'In October 1702, I caused the filth or dirt of the gutters, when there was no water there, and the dirt was quite dry, to be gathered together, and took about a tea-cupful of the same, and put it into a paper upon my desk, since which time I have often taken a little thereof, and poured upon it boiled water, after it had stood till it was cold, to the end that I might obviate any objection that should be made as if there were living creatures in that water. These animalcula, when the water runs off them or dries away, contract their bodies into a globular or oval figure. After the above-mentioned dry substance had lain near twenty-one months in the paper, I put into a glass tube, of an inch diameter, the remainder of what I had by me, and poured upon it boiled rain-water after it was almost cold, and then immediately viewed the smallest parts of it, particularly that which subsided leisurely to the bottom, and observed a great many round particles, most of which were reddish, and they were certainly animalcula; and some hours after, I discovered a few that had opened or unfolded their bodies, swimming through the water; and a great many others that had not unfolded themselves, were sunk to the bottom, some of which had holes in their bodies; from whence I concluded that the little creature called the mite had been in the paper, and preyed upon the aforesaid animalcula.

'The next day I saw three particular animalcula swimming through the water, the smallest of which was a hundred times smaller than the above said animalcula.

'Now,' continues this quaint observer, 'ought we not to be astonished to find that these small insects can lie twenty-one months dry, and yet live, and as soon as ever they are put into water fall a-swimming, or fastening the hinder parts of their bodies to the glass, and then produce the wheels just as if they had never wanted water. . . . Some of the bodies of these animalcula were so strongly dried up, that we could see the wrinkles in them, and they were of a reddish colour; a few others were so transparent that if you held them up between your eye and the light, you might move your fingers behind them, and see the motion through their bodies.'

What Leeuwenhoek saw at the beginning of the last century, naturalists have frequently observed since; and we thus find that the peculiarities of these animalcules' constitution enable them to be dried, to be kept in this dried state for many months, and more wonderful still, on the addition of a little water to resume all the functions of life with renewed vigour. Nor is this all. They may be exposed to a temperature of 104° F. without suffering any apparent harm; whilst dried animalcules have been subjected to a heat of 144° F. with the same result. Dr Carpenter tells us that he dried and revived some of these animalcules six times in succession; and Professor Owen remarks that an animalcule was revived at Freiburg after having remained four years in a mummy-like condition.

These astonishing facts find a parallelism in the case of the seeds of plants, which have been known in some cases to have remained in a dry and parched state for lengthened periods of years—cases well exemplified by the mummy-wheat and mummy-peas obtained from Egyptian tombs, and by the reappearance of plants after an interval of many centuries. Certain seeds in one case, indeed, were buried under the refuse and slag of some mines in Greece, which had not been worked since the classical period, and yet on the removal of the *débris* sprang up into full vitality.

The seed of the plant and the dried-up wheel-animalcule are said to exist in a 'dormant' state, but this statement brings us no nearer than before to the actual explanation of the why and wherefore of the curious conditions the animal and plant are able to assume. One fact at anyrate is plain: the dried animalcule is certainly not *dead;* it is simply in a state of 'suspended animation,' similar to that exhibited by the half-drowned man or dog. We may *revive* an animal or plant in which life still exists; but this process, it need hardly be said, is not equivalent to *revitalising* or *revivifying* a dead being. But regard the present subject in whatever light we may, its consideration seems to add one more to the many puzzles which compass our own existence, and which invest the life even of a humble animalcule with an interest which none but the wilfully blind and ignorant can neglect and despise. The study of such subjects, indeed, alone can save us from the impeachment, patiently borne by no earnest mind, of

Moving about in worlds not realised.

LIFE IN THE DEPTHS.

WITH the history of those minute animals named *Foraminifera*, which form in all existing oceans a thick deposit of limy matter, most of our readers will doubtless be familiar. The accounts of deep-sea dredging expeditions have been so frequently alluded to and commented upon in newspapers and magazines, and the recent voyage of H.M.S. *Challenger* has been so often described in the same media, that every one must have heard something of those minute organisms which, year by year, increase in importance in the eyes of the geologist, and of his scientific brother, the natural historian. With the history of the Foraminifera and their neighbours, much that is both puzzling and interesting is bound up; and it may prove instructive if we glance, even in a superficial fashion, at the general relations of these curious little organisms.

If we regard their position in the present system of zoological classification, we shall find them to be placed by naturalists in the lowest sub-kingdom or great primary group of the animal series. To this great group of animals the name of *Protozoa* has been given; and if we wish to obtain a general idea of the nature of our Foraminifera, we should most readily obtain that idea by defining them as usually minute animals, inclosing their bodies in shells or more properly *tests*, composed usually of lime, or sometimes, but more rarely, formed of grains of sand cemented together; whilst some may be protected by a horny covering. They are thus 'shelled' animals, in the sense, at anyrate, that they possess a covering resembling the structure we ordinarily denominate a 'shell;' and it is this shell, or more clearly,

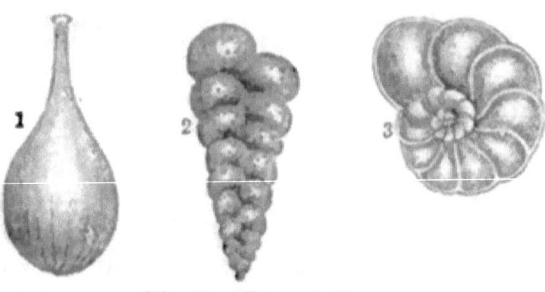

Fig. 5.—Foraminifera:
1. Lagena or 'Flask-animalcule;' 2. Textilaria; 3. Operculina.

the fact of their possessing hard parts, which has brought the Foraminifera so prominently under the notice of the geologist. If we wish to procure them for investigation, we may find them in abundance in all existing seas. In almost every region, and from very great depths, the dredge comes up loaded with these organisms, which constitute the greater bulk of the peculiar chalky 'ooze,' or mud, so familiar to the investigator of the deep seas. We may obtain these shells upon our own shores amongst the sand; or we may find some species at low-water mark living on the tangle-fronds that border the rocks and stones. And in the rock-formations of the earth, which represent in themselves the worlds of the past, we may also find these organisms in plenty. From the earliest or oldest rocks in which the fossilised remains of living things have been found, we may obtain Foraminifera; and they range through the stratified rocks from these oldest beds to the present day. Sometimes, and in particular rock-systems, as we shall presently notice, they attain a development which fairly startles us by its immensity; and it would thus seem that, as far as regards their distribution, whether in the past or in the existing world, the Foraminifera are almost ubiquitous. If we regard them as a class of living or existent organisms, we may no less be struck by the variety of form and shape which marks these forms. Thus we may find them appearing as minute spherical bodies—such as *Orbulina*, so named from its rounded form. Some appear to possess equally simple forms, but exhibit a flask or bottle-like shape—such as *Lagena*—the 'flask-animalcule' (fig. 5, 1) of the microscopist. Sometimes we observe this simplicity of form to be exchanged for shapes of compound nature. *Nodosaria* appears before us as a straight-shelled form (fig. 7, *b*), looking very much like a beaded rod. Some have shells coiled up in a spiral, *d*, like the well-known nautilus shell; whilst others, such as the familiar *Globigerina*, possess the segments of the shell disposed in an irregular manner. Lastly, some Foraminifera, of which the famous *Nummulites* (fig. 9)—so named from their resemblance in shape to coins—may be cited as examples, exhibit shells of a still more complicated type of structure; and we know of other forms, to be hereafter noted, which existed in large reef-like masses, and which present apparently at first sight little resemblance to their simpler neighbours.

We may now, however, glance at the animals which inhabit and manufacture these shells. Primarily, then, we note the amazingly simple nature of the living organisms which are the actual *Foraminifera;* since we must certainly give the title itself to the living tenants, and not to the mere houses or shells. Each living Foraminifer consists of a simple, minute speck of that peculiar substance named *sarcode* or *protoplasm*, of which, in its simple and primitive state, and as illustrated by the animalcules, the bodies of the lower animals and plants are composed. True it is, however, that the bodies of all animals and plants high and low, man himself included, are composed of this same essential 'matter of life.' And biology gives the death-blow to pride of heart, when it truly asserts that the bodies of man and the monad are essentially made of the same material; which, in the case of the higher animals, has, as it were, been elaborated by the type and extent of development from the primitive material of the lower being, to form bodies of more or less complicated kind.

The sarcode of the Foraminifera makes its appearance as a jelly-like substance, of reddish colour, devoid of all elaboration,* and exhibiting no traces of organisation or structures of any kind. And we thus perceive that a minute speck of this organic matter is sufficient to constitute, of itself, a truly living being; which, however apparently simple its structure and lowly its place in the scale of being, yet presents problems which mock the efforts of the most advanced science in its endeavour to solve them. Such a being not only eats and nourishes itself, and performs all the functions of its simple life, but, as we have already seen, may form a complicated shell. The lime of the surrounding water is thus laid hold of and secreted by the living matter, and in due time appears transformed and built up into the shape of the shell. Thus effectually, silently, and unaided, does the humble animalcule accomplish a work which would tax human energies beyond their utmost powers; and thus do we perceive, even in such a

* The most recent researches, by Schulze and Hertwig of Germany, would appear to shew that the protoplasm of these minute organisms is not of so elementary a character as has been hitherto supposed.

superficial study as the present, the grand distinction between the living and non-living world. An implied power of action, bringing into use the surrounding circumstances of its life, characterises the living being, wherever and however it exists. And in the manufacture of its shell, each tiny Foraminifer possesses this wondrous power in common with the highest being that avails itself of its will and instincts to seek its daily food.

The name 'Foraminifera' is derived from the Latin *foramen*, a 'hole' or 'aperture,' and is applied to these organisms in allusion to the apertures which usually exist in their shells, and through which processes of the soft living sarcode-matter of their bodies are pushed out. These processes are named by the naturalist *pseudopodia* ('false feet'), and in Foraminifera they are of long, delicate, and interlacing kind (fig. 6). Microscopic observers have been able to detect that through the interlacing network of these pseudopodia, a circulation of the granules or solid particles of the sarcode is continually being carried on. Doubtless this circulation is connected with the nourishment of the living matter; and in some complicated forms of the Foraminifera its course and nature may become of a more intricate kind than that of the simple forms. In Foraminifera which possess shells of a porcelain-like structure, no *foramina* or apertures exist in the walls of the shells; the filaments or pseudopodia being protruded from the mouth extremity of the shell. In the other chief variety of shell, which is of glassy structure and is accordingly named 'vitreous,' the pseudopodia are emitted through numerous holes in the walls of the shell. The obvious uses of these filaments are those of serving for the prehension of food-particles, and for the purpose of locomotion. Particles of nutriment are seized by them and drawn into the interior of the body, whilst by their contraction and extension the animals also move about.

We have already spoken of the Foraminifera as exhibiting a division into 'simple' and 'compound' forms. It is interesting to note that the compound shells, and of course their included living parts also, are derived from the simple forms. In other words, each compound Foraminifer begins life as a simple being (fig. 7, a), and attains its compound form by a process of literal budding. New segments are budded out from the single and primary one; and according

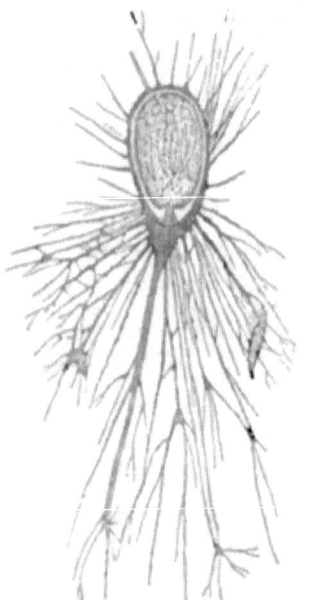

Fig. 6.—Gromia:
A Foraminifer represented in its living state, and shewing the living matter extruded from the shell.

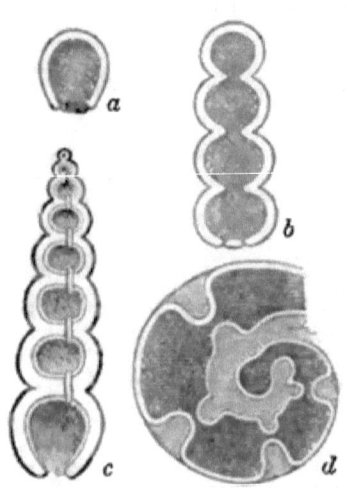

Fig. 7.—Formation of Foraminiferous Shells:

a, a simple one-chambered shell; b, a compound shell formed from a, by budding proceeding in a straight line; c, *Marginulina*, a straight conical shell; d, a shell formed by a spiral process of budding.

as the budding proceeds in a straight line (fig. 7, b, c), in a spiral, d, or in other directions, so we have our straight, spiral, or other forms of shell produced. A compound shell thus consists of many chambers, filled with living sarcode; each chamber containing as it were a single individual of this compound organism; whilst all the chambers communicate with each other, and the sarcode is thus continuous throughout the entire shell.

Having briefly glanced at the structure and living relationships of the Foraminifera, we may next note the interesting facts which the naturalist and geologist have to tell us respecting their distribution in *space* and *time* respectively. In the beds of all our existing oceans, as we have already remarked, we find the Foraminifera to form a thick layer or deposit, which, as time rolls onwards, tends to become of greater extent. Much discussion has taken place amongst naturalists as to the exact *habitat* of the Foraminifera, and as to whether they inhabit the deeper or more superficial waters of the sea. But as far as the question has been authoritatively examined, the evidence would seem to shew that certain species inhabit deep waters, whilst others prefer the upper strata of sea. Thus the most familiar species of Foraminifera, namely the Globigerina, is a surface-living organism, and is obtained by the towing-net in plenty in a living state. Recent deep-sea dredging expeditions have thrown much light on many of the conditions of modern foraminiferal life; whilst dredging and sounding experiments first revealed to us the fact that life was represented at all in the sea-depths; and it is also extremely interesting to note that these beings were among the first-discovered inhabitants of the great depths of the ocean. The areas of the ocean traversed by warm currents are those in which the deposits of Foraminifera occur in their most typical development. And thus the deep-sea ooze consists in greater part of the remains of modern Foraminifera, which, as we shall presently note, become related in a striking manner to the development and life of their ancient representatives.

The *Challenger* expedition, in particular, has thrown a vast flood of light on the distribution of the Foraminifera in existing seas, and perhaps no more interesting point in their history can be mentioned, than that which relates to the alterations produced on their shells at certain depths in the ocean. At depths of 2000 fathoms, the chalky ooze or mud, formed of the shells and *débris* of Foraminifera, retains its common and usual characters. When brought up from that depth, and examined by the microscope, we can recognise the Globigerinæ and their neighbours; the spines which the Globigerina shells possess when found alive at the surface of the water, having been broken off. In depths below 2000 fathoms, however, the chalky ooze and its Foraminiferous shells appear to undergo a singular change. At 2500 fathoms depth, for example, the bed of the ocean is found to be covered by a red clay, which has apparently taken the place of the white ooze of the upper waters. This red clay further exhibits in its chemical composition a remarkable likeness to common clay; the limy matter of the chalky ooze having apparently been removed or altered. The explanation of this singular disappearance of foraminiferal shells from depths below 2000 fathoms has been supposed to rest upon the fact that a peculiar chemical action is exerted upon the limy matter of the shells, when these sink below the limit just mentioned. This action, probably produced by the excess of carbonic acid gas, dissolves the shells, and thus completely destroys their identity. A certain proportion of the red clay, which consists of alumina and iron, is probably formed from materials derived from the earth's crust; but there seems every reason to believe that it in greater part represents the 'ash' or residue of the foraminiferous shells which have been dissolved by chemical action. Such a study, it may lastly be noted, helps the geologist to understand how it is, that in rocks containing abundant Foraminifera and other fossils, large tracts, barren and destitute of the remains of living beings, should occur. The

LIFE IN THE DEPTHS. 33

petrified red clay would thus present us with a rock barren of life, and yet allied to and continuous with rocks, which, as represented by the petrified chalk-ooze, would contain abundance of fossils.

If we now turn to the geological history of our Foraminifera, we find these forms representing the first traces of animal life known to the geologist. In the Laurentian rocks of Canada, and lying towards the base of that series of rock-formations, is a deposit of mineral matter named 'Serpentine Limestone.' This deposit consists of layers of chalky material, arranged alternately with bands of a mineral named serpentine or 'silicate of magnesia' (fig. 8, A); and when the limestone layers are microscopically examined,

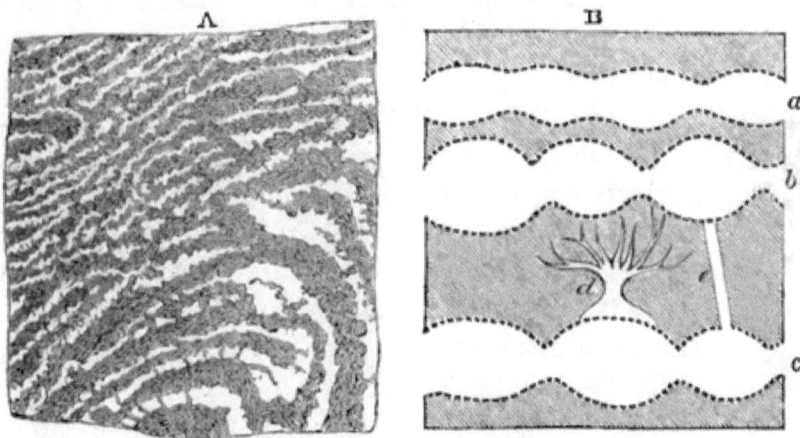

Fig. 8.—The Dawn of Life Animalcule or *Eozoön*:
A, portion of *Eozoön* (natural size); B, diagram of its structure: shewing *a*, *b*, *c*, three tiers of chambers; *d*, tubes contained in the shell-wall; and *e*, passage of communication between one tier and another.

they are found to present a structure of distinctly *organic* nature—that is, indicating their origin from living beings. To this organic structure the name of *Eozoön Canadense*, or the 'Dawn of Life Animalcule,' has been given. And from a close examination of its structure, we are led to believe that the Eozoön was a Foraminifer which grew in immense reef-like masses, comparable, in their mode of growth at least, to the modern coral-reefs. Apparently the shell of Eozoön consisted of a series of chambers arranged in vertical tiers (fig. 8, B); the chambers themselves being partly partitioned off or divided, after the fashion we have already noted in existing compound Foraminifera; whilst communications existed between the various tiers, *e*, so that the sarcode or living matter of this great colony was made continuous throughout its extent. A peculiar system of tubes or canals, *d*, has been discovered branching out within the layers of the shell, and this 'canal system' has its representative in the shells of living Foraminifera also.

Eozoön, which also occurs in formations in Ireland and in Central Europe, thus represents the oldest traces of life with which we are acquainted; and it is certainly bewildering to think of the immensity of the periods of time which have elapsed since the existence of the primitive ocean in which the living Eozoön grew and propagated its reef-like masses. Geology, which has no historical or absolute chronology, refuses to set any limit in years to that time, and the question is one which perhaps, after all, and like the idea of space, is best left unanswered as belonging to the infinite itself. Eozoön has one or two representatives in existing seas—at least in the mode of its growth. Two examples especially—*Carpenteria*, named after Dr W. B. Carpenter, whose researches into foraminiferal life have been of the most complete character, and **Polytrema**, a branching Foraminifer—exemplify this condition of aggregation in reef-like colonies or masses.

Passing upwards in the scale of rock-formations, and guided by the age of the deposits, we meet with many examples of fossil Foraminifera. The Silurian rocks contain these shells very plentifully in some localities; and one species—the well-known *Fusulina*—forms by its extreme development in numbers, whole beds of limestone belonging to the Carboniferous or coal epoch in Russia, and other parts of Europe. When we arrive at the Cretaceous or Chalk rocks, we find a stage in which foraminiferal life must have existed in greatest luxuriance. The enormous cliffs of the true or White chalk, nowhere seen to greater advantage than in Albion itself, are composed almost entirely of foraminiferous shells, many species of which are identical with those of our existing seas. The white cliffs of the south of England represent merely huge monuments of foraminiferal life, and carry our thoughts backward to an old ocean in which a deposit similar in kind to that taking place in our existing oceans, but of vastly greater extent, occurred. Thus we read the past by our knowledge of the present; and if we turn to the existing state of affairs, we could readily imagine that were the bed of our present ocean elevated and petrified, the foraminiferal deposit of to-day would come to resemble the ancient chalk. If we moisten a piece of chalk in water, separate out the particles, and microscopically examine them, we should find our fossilised chalk to consist of foraminiferal shells and fragments, exactly corresponding to such a deposit as we might prepare by similarly treating some ooze from our deep-sea dredge. In particular, the *Globigerinæ* would be seen in the chalk, indistinguishable from those which we may procure in profusion from existing seas; and *Rotalia*, *Textularia*, &c., well known as living forms, are also to be seen represented by species in the chalk. So strongly have these facts become impressed on the minds of some geologists, that it has been asserted that we are still living in the Cretaceous or Chalk age; although as viewed by other authorities, the remark is true in a restricted sense only, if indeed its truth can be admitted at all.

Approaching relatively nearer to existing times, and coming to the newer rocks, we find in the Eocene rocks a remarkable profusion of foraminiferal life. Here are found the Nummulites, large forms, which may sometimes be found to measure three inches in circumference. These organisms form the rocks known by the name of the Nummulitic Limestone, which runs from the Pyrenees and Alps to the Carpathian Mountains; which is found in North Africa, and may be traced from Egypt to Asia Minor, and onwards through Persia, by way of Bagdad, to the mouths of the Indus. This limestone is also found—we quote from Lyell—in Cutch, in the mountains between Scinde and Persia, and it may be followed out eastwards into India, Eastern Bengal, and to the Chinese frontiers. In thickness, the Nummulitic limestone sometimes rivals the older Chalk, and attains a depth of several thousands of feet. This great deposit is literally composed of Nummulites and of their *débris*, massed together to form a solid rock; and from this deposit, the stone of which the Pyramids are built was quarried. Thus in the history of the materials of which these strange edifices are composed, no little share of romance and wonder may also be said to enter. The Nummulites consisted each of a complicated series of chambers or segments, developed in a spiral manner so as to form a flat coin-shaped structure, which exhibits a complicated arrangement of its internal parts, particularly in the development of the 'canal system,' already alluded to as occurring in Eozoön. In the Eocene Rocks we also meet with the Miliolite limestone, a deposit which forms the 'basin' in which Paris lies, and from the materials of which the houses of that city have been built.

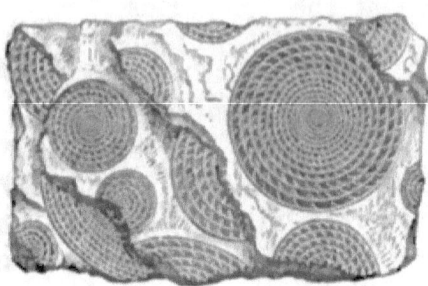

Fig. 9.—Nummulites.

With living *Miliolæ* we are well acquainted, and these forms, by their great development in the Eocene period, have thus contributed to give to man, by the aggregation of their shells, the material for beautifying his cities and dwelling-places.

From the Eocene rocks to the deposits of our own day is a transition of comparatively slight extent, and we may therefore, with the history of the Eocene period, take leave of the Foraminifera. We may, however, conclude our study of these little organisms by a glance at some of their near neighbours, known as *Radiolaria*. The latter possess flinty shells, and in this respect differ from the Foraminifera; whilst one can observe no structures exhibiting stricter mathematical or more regular outlines, or shewing forms of greater beauty, than some of these minute flinty shells. The Radiolarians live at all depths in the sea, and

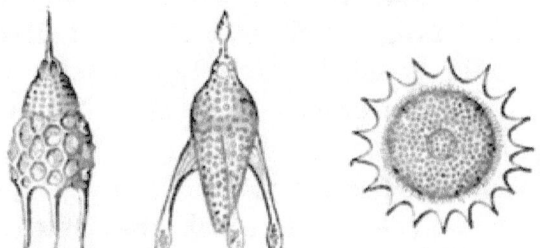

Fig. 10.—Various forms of Radiolarians.

occur plentifully as fossils in certain rock-formations, such as the wrongly-named 'Infusorial Earth' of Barbadoes. It forms not the least puzzling point in the history of these animalcules to account for the power possessed by one little speck of protoplasm of manufacturing a limy shell from the water of the ocean, whilst a neighbour and almost identical particle secretes a flinty shell. This power of selecting materials is in fact one of the unexplained and truly difficult points in the history of living beings.

It is not always from the great things of science, or,

indeed, of life itself, that the most valuable lessons are to be derived; and the history of the Foraminifera may tend to shew how great a fund of thought and information lies hidden in the consideration of a group of organisms which many might deem too insignificant to merit much attention. Such studies also afford a strong argument in favour of natural science forming an element in the education of the young. The habits of observation and of regularity induced by the study of natural objects, cannot be too highly valued or over-estimated as serving to train the youth of both sexes in the use of thorough method. Any one who daily sees the want of order and punctuality in the business of ordinary life, cannot but become a sincere advocate of any branch of education, which, in a pleasant and instructive manner, may lead to the formation of habits of method and order in the young. And such studies, moreover, are also useful in encouraging a love for the beautiful and the true. It should form no insignificant part of the studies of youth, that they should be taught innately to admire whatever of beauty and good this world can be shewn to contain; for, from their appreciation of such things, will in due time follow a sincere belief in the precepts of religion, and in the elements of morality—studies which are limited by no day or age, but which hold good for all time.

CONCERNING SEA-ANEMONES.

And here were coral-bowers,
And grots of madrepores,
And banks of sponge, as soft and fair to eye
As e'er was mossy bed
Whereon the wood-nymphs lie
With languid limbs in summer's sultry hours.
Here, too, were living flowers,
Which, like a bud compacted,
Their purple cups contracted ;
And now, in open blossom spread,
Stretched like green anthers many a seeking head.

SUCH a description, poetic and ethereal as it may seem, is not by any means overstrained, if applied to the scene which meets the observer's eye as he peers into the clear depths of some rock-pool, which lies embowered in some rocky niche, and retains its calm waters until the next inflow of the tide. Such an observer may be said literally to gaze upon a world in miniature, teeming with life of almost every kind and grade. The dark waving masses of tangle and other sea-weeds may fitly represent the forest trees, and these are surrounded by sea-weeds of less sombre hue, presenting tints often of gay and pleasing kind. Each frond of sea-weed seems to bear its quota of life. Swift-swimming prawns with clear, glassy bodies, dart in and out of the rocky grots like spectres. Here and there the slower crab sidles along with awkward gait; and even the finny tribes may be represented by some of the

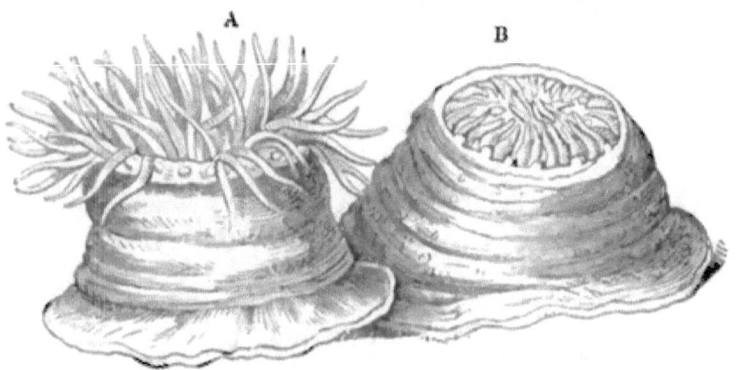

Fig. 11.—Sea-Anemones:
Actinia mesembryanthemum, the ' Beadlet,' or Common Sea-anemone:
A, expanded; and B, contracted.

smaller fry which find in the rock-pools at once a sea of fair dimensions, and a home secure from the turmoil of the ocean beyond.

Ensconced securely on rocky ledges, and serving to render their surroundings gay and varied by the beauty of their dress, the sea-anemones in such a scene present perhaps the objects of greatest interest to the naturalist. Southey's epithet of 'living flowers' conveys a most appropriate idea of the general appearance of the anemones of the sea; and as we may find literal forests of animals represented by the zoophytes, so we may behold the gay parterres and flower-plots of marine gardens in the commonest of the tenants of our rock-pools. Thus zoology gives some countenance to the poet's ideas that

> Seas have—
> As well as earth—vines, roses, nettles, melons,
> Mushrooms, pinks, gilliflowers, and many millions
> Of other plants, more rare, more strange than these,
> As very fishes, living in the seas.

The reader who has visited an aquarium must have regarded with wonder and interest the literal zoological flower-show which the anemone-tank may be said to represent. Apparently rooted and attached to the rocks, we see beings of varied hues and shapes, but all agreeing in possessing soft fleshy bodies, crowned by circlets of delicate filaments—the 'petals' of our animal-flowers—and in possessing a central opening in the middle of the petals. This general idea of sea-anemone structure thus introduces us to a cylindrical body, fixed by one extremity, and bearing at the free or opposite end a mouth, surrounded by numerous feelers or tentacles. The appearance of these veritable animals is thus truly flower-like; and it is perhaps not surprising to learn that their truly animal nature was first discovered only some hundred and fifty years ago. Prior to that period no doubt existed in the minds of naturalists that the sea-anemones and their allies were marine plants; sensitive, no doubt, but plants nevertheless. The effects of this belief were well exemplified by the publication in 1706 of the researches of Count Marsigli, a celebrated French naturalist, who described and figured the animals which manufacture the red coral as true flowers, each possessing eight sensitive 'petals.' And as the coral-animals represent literal colonies of sea-anemones, Marsigli's ideas regarding the coral-'flowers' applied equally well to the animals under discussion. Some twenty years afterwards, one of Marsigli's pupils, Peysonnel by name, being sent to study the supposed coral-plants on the coasts of the Mediterranean Sea, discovered that the coral-'flowers' were in reality little animals; and as the sea-anemones in 1710 had been described as animals, Peysonnel compared the coral-animals, and very justly as modern zoology shews, to these more familiar forms. We need not further trace out the history of the disbelief in Peysonnel's researches which followed, save to remark, that the subsequent progress of investigation amply vindicated the correctness of his views, and elevated the sea-anemones and their neighbours for ever above the level of the plant-world. The relationship between the corals and the sea-anemones, as already remarked, is of the closest kind; and no better preparation for the study of the coral-polypes could be had, than by making the acquaintance of their more familiar representatives—the 'living flowers' of the poet.

When we examine a piece of red coral in its living state, we thus find that the hard coral-substance exists within the soft parts; the living animals covering the central coral as the bark invests a tree. This living bark consists of a soft skin, imbedded in which we find numerous little animals, each possessing a central mouth surrounded by eight fringed tentacles. Each coral-polype, in fact, not only resembles a little sea-anemone in its outward appearance, but corresponds with the latter in all essential details of structure. The chief differences between the sea-anemones and the coral-polypes consist, firstly, in the power possessed by the coral-animals of producing great colonies by a literal process of 'budding;' and secondly, in the manufacture of the limy skeleton we term the 'coral'—the sea-anemones possessing no hard parts whatever. Thus we may accept the anemone as the type and representative of the great groups of coral-producing animals, as far as the nature of the living parts of the latter is concerned. And we also may note how erroneous ideas are spread regarding the nature of the coral-polypes, when they are named coral 'insects' and the like. The term 'insects' as applied to the coral-polypes is seen to be quite inappropriate; and it is to be regretted that it is used in any case to denote animals, which possess no relations whatever with the insects truly so called.

With regard to the structure of the anemones, the naturalist finds that they possess bodies of very simple kind. The rounded body-wall encloses a space of similar shape; and the central mouth (fig. 12, *m*) opening on the upper surface of the body, is found to lead directly into a stomach-sac,

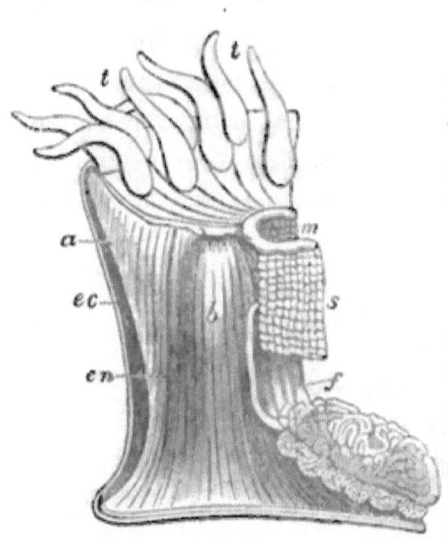

Fig. 12.—Dissection of Sea-anemone: *a*, muscular layer of body-wall; *b*, mesentery or partition; *ec, en*, layers of the body-wall; *f*, body-cavity; *m*, mouth; *s*, stomach-sac; *t, t*, tentacles.

s, which, curiously enough, can be compared to nothing liker than a pocket with the bottom cut out. This stomach-sac hangs down into the interior of the body-space, *f*, and communicates freely below with the latter. It is kept in its place by a series of partitions, *b*, which spring from the body-wall and run inwards to become attached to the stomach. Thus it will readily be understood that, if we cut the sea-anemone's body across, we should find it to be composed of two tubes; the outer tube being represented by the wall of the body, and the inner tube by the stomach-sac; whilst the intervening space would be divided into a number of compartments by the partitions already mentioned.

The tentacles, *t*, which surround the mouth are simply hollow upward growths of the body. Each tentacle is a simple tube, possessing a sucker perforated by a hole at its tip. The functions of these structures may be readily ascertained if any substance adapted to serve as food is brought in contact with them. In such a case, exemplified, for example, by the act of an unfortunate stray crab coming in contact with the tentacles, these organs are seen to seize the morsel by their sucker-like tips and to drag it by their contraction to the mouth, within which the morsel disappears; whilst the tentacles themselves are tucked in within the same wide aperture. That the tentacles, as just remarked, are simple extensions of the body itself may be proved by any one who causes an anemone to contract itself with some degree of force. In the latter case, the water contained within the interior of the body will be seen to be forced out at the tips of the tentacles, as well as by the mouth.

The observation of an anemone's daily life reveals a series of highly instructive and interesting details. That the animals are highly sensitive, is not only a familiar but a most important observation. We see this sensibility to outward impressions when we touch one of the tentacles. The slightest touch is sometimes sufficient to cause the whole of the tentacles to be withdrawn and folded inwards, whilst the animal comes to look like a conical mass of jelly, coloured of varying hues, but utterly unlike the graceful being which represented the placid undisturbed anemone. After an interval has been allowed to elapse, the tentacles will be gradually expanded, and once more with 'unruffled plumes,' the anemone will throw its beauty open for admiration. Southey well describes the expansion of the sea-anemones, when he speaks of

> The living flower that, rooted to the rock,
> Late from the thinner element
> Shrunk down within its purple stem to sleep,
> Now feels the water, and again
> Awakening, blossoms out
> All its green anther necks.

The old proverb that 'familiarity breeds contempt,' would seem in a manner to be illustrated by the difference observed between the sensitiveness of sea-anemones in the rock-pools, and of those which have been kept in an aquarium, and which have been handled more or less frequently. The author well remembers the case of two anemones which he possessed and kept in a small aquarium for about two years. At first, the anemones were highly sensitive to the slightest touch; but as time passed, they became, as it would appear, more familiar, and might sometimes be freely touched without contracting beyond a few of the tentacles. It is quite possible, therefore, that custom and habit may render even these lower organisms less susceptible to influences and impressions, just as these conditions powerfully operate in inducing a sense of toleration in man. The anemones possess, however, more acute sensibilities than those evoked by touch. The sudden interruption of the light which has been playing upon them, as when the sunlight has been obscured by a passing cloud, will cause them to retract their tentacles. They are thus sensitive to light and darkness, and this fact may be explained by the knowledge that around the mouth we may find a row of little colour-spots, which we must regard as rudimentary or simple eyes.

But unquestionably the most mysterious feature in the sea-anemone's sensitiveness consists in the fact, that these animals thus appear to feel, without possessing any means for exercising that sense. No traces of nerves are to be discovered in these animals after the most minute examination of their tissues by the searching powers of the microscope. And hence physiologists may find in the sea-anemones a text for a discussion upon the important question of 'nerves and no nerves, or the art of feeling.' The anemones are not singular in respect that they exhibit sensitiveness in the absence of nerves. Still lower animals, such as the zoophytes and their neighbours, the jelly-fishes, the hydræ of our ponds, and other organisms, appear to feel and to exhibit irritation, without possessing a nervous system as part of their organisation. And we know that some plants —of which the best known, perhaps, is the Sensitive Plant, which droops **its leaves on being touched**—appear to be as truly sensitive as the anemones and **their** neighbours. Thus, firstly, sensation is **not confined to the animal world; and** secondly, in plants, **and in many lower animals, sensitiveness** is thus noted to **exist independently of the presence of any distinct nervous system.**

Naturalists are led to explain this apparent discrepancy between the powers exhibited by living **beings and the** structures with which they are provided, on the supposition that the tissues of the lower animals **possess a** general or diffused sensitiveness which represents the definite **nervous powers of** higher **animals.** The expectation **that every action of an animal or plant must be performed by** distinct **organs and parts, gave rise to an** idea formerly **much in vogue, that life itself was dependent on the presence of organs and structures; or in other words, that an** animal lived because it was organised. **So far is this from** the truth, that we know of **lower animals and lower** plants which live without possessing any distinct organs or structures at all. A simple speck of living jelly-like matter, such as the amœba, is seen to constitute a living animalcule, which lives **as** perfectly in its own way as the highest of animals. And it is therefore, no more wonderful to think of a sea-anemone exercising feeling without **nerves,** than **to reflect that an** animalcule lives, eats, **and moves without having any distinct** apparatus to perform **these functions. In short,** we must guard against regarding **animals as mere machines;** since we frequently find in living beings **a great** disparity **between** the **acts they** perform, and the machinery which does the work.

Similarly in the jelly-fishes, which are very near neighbours of the sea-anemones, no nerves have been found; but very striking proofs of the exact nature of the sensitiveness in these animals have quite recently been afforded. When the margin of a jelly-fish was touched, the central mouth, which is borne on a stalk, was moved towards the point which was touched. Time after time, the stalked mouth indicated by its motions, its sensitiveness to the exact point of touch. When on the other hand, a horizontal incision was made in the soft body of the jelly-fish, and the body was touched

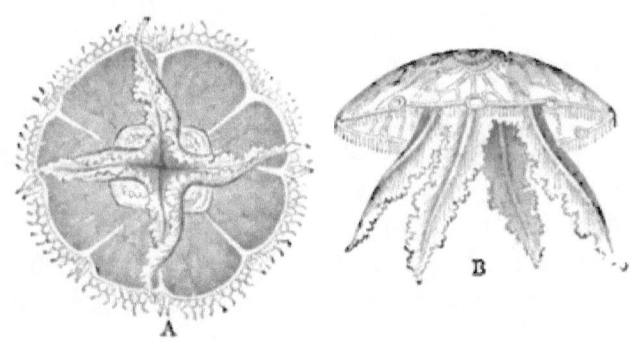

Fig. 13.—Medusa or Jelly-fish:
A, under surface, shewing the mouth in the centre, surrounded by the tentacles; B, side-view, shewing the tentacula hanging down in their natural position.

below the cut, the mouth moved in an irregular manner, as if in quest of information as to the point touched. In this case, the horizontal cut had destroyed the nervous connection between the mouth and the part touched, just as the division of a telegraph-wire interrupts the communication between the stations it connects. And it would thus appear that, although no distinct nerves can be shewn to exist in the jelly-fish, there must nevertheless exist very definite lines or tracks along which nervous impulses travel; since, when these lines are interrupted or broken, the nervous impulses are correspondingly affected. Like the anemones, the jelly-fishes possess little eye-spots, and they further resemble their fixed companions in that they are sensitive to light. On a jelly-fish being suddenly exposed to light, contraction follows; and it has further been determined that the margin of the bell-shaped body where the colour-spots exist, is that portion in which the sensitiveness to light exists; since, when this portion of the animal was cut off, the body became non-sensitive to light, but the amputated portion continued to be as sensitive as when it formed part of the jelly-fish body.

Returning to our sea-anemones from this digression—which, however, has served to explain the nervous acts of these forms—we may next note that these animals are not permanently rooted and fixed to their rocky abodes. They possess the power of moving slowly along by contracting and expanding the fleshy base or root of the body. On the glass of an aquarium the track of a travelling anemone may sometimes be as plainly seen as the mark left by a crawling snail; and several authors have described them as also moving about mouth downwards by using the tentacles as feet.

The brilliant hues of the anemones are developed within the tissues or membranes of which their bodies are composed. The brilliancy of these colours is only equalled by their variety; since they may be met with of all tints, from a pure white to a deep crimson; whilst variegations of orange, blue, and green are by no means uncommon. As every one knows, the presence of light is usually regarded as being necessary for the development of colour both in animals and plants. It is exceedingly interesting, therefore, to note that anemones obtained from deep water, and which live in abysses to which no light can penetrate, have been found to exhibit as bright hues as do their neighbours of the coasts. The *Challenger* expedition brought to light, from a depth of 600 and 2750 fathoms

CONCERNING SEA-ANEMONES.

respectively, anemones which exhibited colours as bright as those of allied species living in shallow waters. At the Philippine Islands one brilliantly-coloured species was found living under a tropical sun, whilst a closely related and as brightly decked neighbour was obtained from a depth of at least three miles—where not only light is absent, but where the temperature is that of freezing-point. These instances serve further to complicate the puzzle of life, and that of the similarity of animals existing under conditions of so varying a nature.

The food of the anemones consists literally of all and sundry that comes in their way. Shells are frequently found amongst the indigestible materials ejected from the mouth, their tenants having been duly digested; whilst crabs form a large proportion of the food. Occasionally, when hungry or in an invalid condition, the anemones may be seen to turn the stomach inside out, so that it appears as a delicate whitish sac protruding from the mouth. The prey appears to be paralysed by the peculiar action of numerous little stinging cells named 'thread-cells,' with which the tentacles and tissues generally of the body are provided.

Perhaps the most curious features in anemone-existence, however, are those brought to light by the study of their reproductive habits and phases. In the little fresh-water *Hydra*, a near relation of the anemones, new individuals may be produced by the artificial division of a single body. The hydra of zoology, in fact, derives its name from a kindred quality to that for which its fabled namesake was so famous. The sea-anemones exhibit similar features in their life-history, since they may be artificially divided in various ways, with the result of forming new individuals through the destruction of one. Thus, if an anemone be cut through from mouth to base, and the two halves kept from uniting, a new anemone will grow out of each half; the cut edges of each half gradually closing up and uniting. If the animal be cut across, a new mouth and tentacles will be developed around the cut edges, and the half left attached to the rock will thus come in time to resemble the former and uninjured state of the animal. When cut obliquely, the animal will gradually repair the injury, so as to restore the symmetry of its parts. These facts have been long known to naturalists, and the author, from his own observations and experiments carried out upon anemones in their native pools, and watched from week to week, as well as upon specimens kept in aquaria, can amply confirm the facts just detailed.

Dr Johnston, author of the *British Zoophytes*, gives a singular instance of the power of the anemone to accommodate itself to unwonted and peculiar conditions of its existence. An anemone, which had swallowed one of the shells of a large Scallop of about the size of an ordinary saucer, was on one occasion brought to this author. 'The shell,' says Dr Johnston, 'fixed within the stomach was so placed as to divide it completely into two halves, so that the body, stretched tensely over, had become thin and flattened like a pancake. All communication between the inferior portion of the stomach and the mouth was of course prevented; yet instead of emaciating and dying of atrophy, the animal had availed itself of what undoubtedly had been a very untoward accident to increase its enjoyment and its chance of double fare. A new mouth, furnished with two rows of numerous tentacula, was opened up on what had been the base, and led to the under stomach; the individual had indeed become a sort of Siamese twin, but with greater intimacy and extent in its unions.'

Even more extraordinary than the preceding facts are

those which tend to shew that fragments of an anemone left attached to a rock after the animal has been rudely pulled away from its attachment, will grow in time into new anemones. These organisms would thus appear to live truly in every part, and to possess a vitality of which man himself might well feel envious. Like most other animals, however, the anemones are capable of producing young from eggs; the eggs frequently undergoing development within the parent-body, and the young anemones in due time escaping into the outer world through the mouth of the parent. Any one who keeps anemones, particularly in summer, in an aquarium, will note that periodically numbers of little anemones, some of them not larger than the heads of pins, appear in the water, and attach themselves to the sides of the glass. These represent the young and rising generation of anemones, and require simply to grow and to develop additional tentacles in order to resemble their parents in every particular.

To speak of the longevity of sea-anemones may seem somewhat ridiculous, but it so happens that we possess the most reliable evidence that these animals, when duly tended and cared for, may live for lengthened periods. The most notable instance of a sea-anemone having attained to an advanced age, is that of 'Grannie,' a specimen of the Common Sea-anemone (*Actinia mesembryanthemum*) of our coasts, which was first taken from its habitat by Sir John Dalzell in 1828. This anemone produced 276 young in a period extending over six years. Between 1828 and 1851, 'Grannie' produced 344 young; and in 1857, gave birth in one night to no less than 240 young. Since 1857, this venerable specimen has been in the possession of Dr M'Bain of Trinity, near Edinburgh, who kindly furnished the details of 'Grannie's' history to the writer. In August 1872, this anemone produced thirty young, and in December of the same year gave birth to nine young. In each year since 1872, at uncertain intervals, 'Grannie' has produced from thirty to forty young anemones. In January 1877 —the last brood, as we write—five young were born; one of the five being four times the size of the others and feeding freely. More interesting still is it to find that one of this anemone's progeny has in turn produced three young anemones, and the title of 'Grannie,' originally bestowed through courtesy, has therefore become a perfectly just cognomen. Some curious facts relative to certain 'messmates' and companions of the anemones may be cited by way of conclusion to their biography. We shall discuss hereafter the details of the curious relationship which seems to exist between a certain species of sea-anemone and a species of hermit-crab. Whilst no less strange and inexplicable is the fact that some large species of sea-anemones, living in tropical seas, afford shelter and refuge to fishes which live in their interior, and swim in and out of the mouths of their hosts at will. Another anemone lives apparently on the best of terms with a little fish; and this fish-guest has been seen to be enclosed within the body of the anemone, when the latter has contracted itself, without suffering any injury. In virtue of what favouring influences the fishes are protected from being digested as food, or what the nature of the association existing between the animals may be, science cannot tell. Such facts tend to shew, at anyrate, that in anemone-existence, as in human life, there are more things than our philosophy can explain.

SEA-EGGS.

THE visitor to the sea-side must frequently in his rambles along the beach have picked up specimens of the curious animals which are popularly known as 'Sea-eggs' and 'Sea-urchins.' The former name is applied to these creatures when they are found cast upon the shore, presenting the appearance of rounded or ball-shaped objects (fig. 14), each enclosed within a hard but brittle limy shell. Whilst the term 'urchin' is given to the same objects when they are seen in their more natural and perfect state, and when the outside of the shell literally bristles with spines. The name 'urchin,' in fact, originally applied to the hedgehog, has been extended to denominate the sea-eggs, from their presenting the spiny appearance so familiarly seen in the common tenant of our woods and hedgerows. Thus the 'sea-egg' is simply the sea-urchin with its spines detached and rubbed off by the unkindly

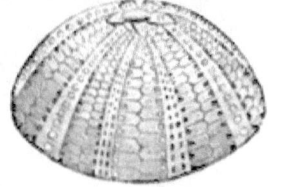
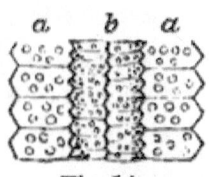

Fig. 14.

Shell of Sea-urchin: *a*, plates without holes; *b*, perforated plates.

force of the waves; and the animal thus popularly designated, is the *Echinus* of the zoologist, and belongs to the large class of animals of which the Star-fishes are well-known representatives.

The entire history of the sea-egg is of so curious a nature, that the most casual reader may well feel interested in the account of the animal's present and past life; whilst the feeling of mere curiosity to know something concerning the history of one of the most 'common objects of the shore,' should prompt every sea-side visitor to make the closer acquaintance of the Echinus.

Suppose that we begin our examination by looking at the hard case or 'shell' in which the soft parts of the animal are enclosed. We find, on referring to the development of the animal, that this 'shell' actually represents the hardened skin of the animal, and that, viewed in this light, it closely corresponds to the shell of the lobster or of the crab. If we break open the shell of the sea-egg, we shall find that it is not composed of one solid mass of limy matter, but is, on the contrary, made up of numerous definite parts, the arrangement of which forms one of the most interesting and notable points in the sea-egg's structure. The shell is flattened at each pole, and we can readily perceive that it is composed of rows of little limy plates, which are disposed in a regular manner from pole to pole, or after the fashion of the meridian lines on a globe. Counting the series of plates, we find the shell to be composed of twenty rows; but we may also perceive a difference between certain of the plates of which the rows are composed (fig. 14). Thus we find two adjoining rows of plates, *b*, which are perforated with holes. The next two rows, *a, a,* are not so perforated; whilst the third two rows possess holes like the first rows. We may, in fact, proceed round the shell and come back to the point at which our examination began, with the result of finding that we may group the whole of the twenty rows of plates of this curious limy box into two sets—those with holes and those without; and we may further discover that there are five double rows of perforated plates, and that these alternate with other five double rows which do not possess holes.

Each little plate of the sea-egg's shell may be most accurately described as being hexagonal or six-sided in form, but this shape may be more or less modified in certain regions of the shell. When the plates increase in size, so as to accommodate themselves to the increasing growth of the animal they serve to protect and of which they form part, the new limy matter is added to the edges of each plate; this new material being formed by a soft membrane which exists at the 'sutures' or joints between them. In all sea-eggs, save a few of the rarer species, the plates are firmly united together, so as to convert the shell into a rigid immovable casing for the animal. In the exceptional cases just alluded to, however, the plates may be less firmly united together, or may be of very thin structure; this conformation giving a certain degree of elasticity, or even conferring a high degree of flexibility upon the shell. The five double rows of the shell which are perforated with holes, it may be remarked, are those, through the apertures in which the small 'tube-feet' of the animal are protruded. And it may also be noted that in some of the sea-eggs these perforated rows do not extend from pole to pole of the shell, as in the common species, but are limited so as to form a rosette-like figure, on the upper surface or at the upper pole of the shell. This modification is well seen in a group of sea-eggs, not uncommon round our coasts, and which are popularly named 'Heart-urchins' from their peculiar shape.

The outside of the shell presents us with some curious features; the zoologist's study leading him thus to note carefully points which an ordinary observer would hardly deem worthy his attention. As we have already remarked, in its natural and perfect state the shell of the sea-egg is covered with spines (fig. 15, *z, z*), which in some cases—as in

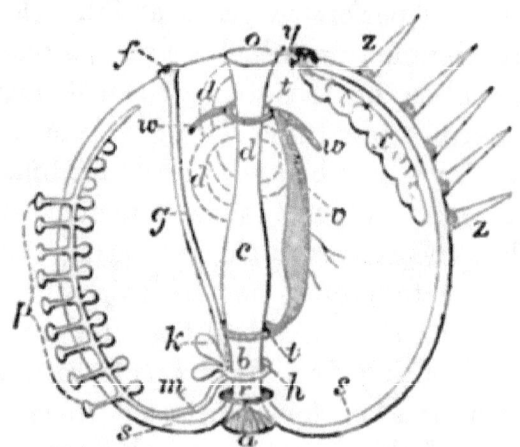

Fig. 15.—Diagram of the Structure of Sea-egg:

a, Mouth, with jaws ('Lantern of Aristotle'); *b*, gullet; *c*, stomach; *d d d*, intestine; *f*, perforated plate; *g*, sand-canal; *h*, ambulacral ring; *k*, vesicles; *m*, an ambulacral tube; *o*, anus; *p*, ambulacra or tube-feet, with their 'contractile vesicles;' *r*, nervous ring surrounding gullet; *s, s*, two nervous trunks, the right terminating above in a small nerve-mass; *t, t*, rings of blood-system, connected by *v*, the contractile heart; *w, w*, two blood-vessels, radiating from the upper ring; *x*, reproductive organ opening at *y*; *z, z*, spines, with their tubercles.

the group of which *Cidaris* is a well-known example—may attain a very large size, and may appear out of all proportion to the shell itself. When we examine the outer surface of the shell, we find it to be thickly studded over with little rounded knobs or 'tubercles,' which are, if anything, most numerous on those parts or rows of plates which are not perforated; and if we carefully study one of the spines we shall find that it is hollowed out or is concave at its base. Clearly, then, the spines are meant to articulate by means of these hollowed or cup-shaped bases with the rounded knobs on the outside of the shell, and in each case a true ball-and-socket joint is thus formed. The spines are thus intended to be moved, and they are not only firmly attached by a ligament or band of fibres to the surfaces of their tubercles, but appear to be moved by special muscles, which form a thin investing layer on the outer surface of the shell. The spines undoubtedly serve as organs of defence, but in some species they are employed as boring-organs to scoop out holes in the sand, or shallow beds in rocks, in which their possessors lie snugly ensconced.

The outer surface of the shell also bears certain very peculiar appendages, known as 'Pedicellariæ.' These little organisms also occur on the outer surface of star-fishes and other members of the sea-egg's class, but regarding their exact nature and functions zoologists are still in doubt. The form of one of these Pedicellarians may be best imagined by figuring to one's self a small or minute stalk attached to the shell, and bearing at its free extremity two or three little jaws, which move actively upon one another, with a snapping motion. These little jaws can be seen to seize upon particles of food, and there is no doubt whatever that they possess a life and vitality independently of the sea-egg or other organism upon which they reside; since their movements are seen to continue after the death of the animal which affords them lodgment. Some naturalists have regarded them as 'peculiarly modified spines;' but the reasons or grounds for this belief are anything but clear, since it is difficult to imagine any reasonable explanation of the means whereby a spine, itself an utterly dead and inert structure, could acquire an active, living, and independent nature. By good authorities, who have not ventured to theorise so boldly, the Pedicellariæ have been regarded as *parasites* of some kind or other; and they may also possibly represent stages in the as yet unknown development of some organisms. Whilst, assuming them to be fully-grown beings, their function, as they exist on the shell of our sea-egg, has been supposed to be that of seizing particles of food, and of removing waste or effete matters.

The internal structure of the sea-egg shews its near relationship with the star-fishes and sea-cucumbers. The mouth (fig. 15, *a*) is the great orifice which exists at the lower pole of the shell; so that, as our sea-egg crawls slowly and mouth downwards over the bed of the sea, or over the floor of its native pools, it can procure food without any very great trouble as regards its conveyance to the mouth. The internal furnishings of the body include a throat, *b*, stomach, *c*, and complete digestive system, *d d d*, along with a very peculiar set of jaws or teeth, lying just within the mouth, the points or tips of the jaws being usually protruded from the mouth-opening. This arrangement of teeth is named the 'Lantern of Aristotle,' and comprises five conical pieces, so arranged together and so provided with muscles, as to be perfectly adapted for bruising the sea-weeds and other kinds of nutriment on which the sea-eggs subsist. Their near neighbours the star-fishes do not possess any teeth, although, curiously enough, the unarmed sea-stars prefer a richer dietary than that which contents their sea-egg neighbours, since they devour large quantities of oysters and other molluscs. Our sea-egg possesses a heart, *v*, for circulating its blood, in the form of a simple tube; and although no distinct breathing-organs are developed, naturalists believe that the blood may be purified by being circulated through a delicate membrane which is named the 'mesentery,' and which serves to suspend and support the digestive organs to the wall of the shell. The fact that this membrane is richly provided with the delicate vibratile filaments known as 'cilia,' and that it is bathed in the sea-water containing oxygen, and which is admitted within the shell, would seem to favour the idea that it constitutes the breathing-organ of these animals.

The sea-egg is not destitute of means for obtaining some degree of knowledge regarding its surroundings; and it obtains its *quantum* of information through the same channel by which man is brought into relation with the world in which he lives—namely, the nervous system. The sea-urchin possesses no structure corresponding to a brain—indeed, in all animals of its nature, the nervous system exists in an unspecialised or indefinite condition. We do not find, in other words, that development and concentration of the parts of the nervous system seen in the highest groups of animals, and which enables these latter to form definite ideas regarding their surroundings, and respecting the world at large. A cord of nervous matter, r, surrounds the gullet of the sea-egg, and from this central portion five great nerves are given off; one nerve-trunk, s, passing along the inner surface of each of the perforated double rows of plates of the shell, to terminate at the upper pole of the body. The only organs of sense developed in the sea-eggs appear to consist of five little 'eyes' of rudimentary nature, each consisting of a little spot of colouring matter and a lens. These eyes are situated on five special plates of the shell, developed at the upper pole or extremity of that structure. We thus remark that the parts of the nervous system, along with other portions of the sea-egg's structure, are developed in a kind of five-membered symmetry—if we may so express it. And it is a singular fact that not only throughout the sea-egg's class do we find the number five to represent the typical arrangement of parts and organs—as is well exemplified in the five rays of the common star-fish—but we also discover that this number is one exceedingly common in the symmetry of flowers. This fact apparently struck an old writer—Sir Thomas Browne—as being a curious and noteworthy feature of the star-fishes and their allies, since we find him inquiring 'Why, among sea-stars, Nature chiefly delighteth in five points?'—although to this suggestive query the learned and eccentric author of the *Religio Medici* gives no exact or satisfactory reply.

The movements of our sea-egg are effected by means of an apparatus which forms one of the most noteworthy parts of its structure. If a star-fish be dropped into a rock-pool, it may be seen to glide slowly but easily over the bottom of the miniature sea in which we have placed it. When we examine the lower surface of this animal's body, we at once perceive the means whereby its movements are performed. For, existing in hundreds, in the deep groove which runs along the under surface of each ray, we see the little tube-feet or *ambulacra*, *p*, each consisting of a little muscular tube, terminated in a sucker-like tip. By means of an apparatus of essentially similar kind, the sea-egg is enabled to crawl slowly over the floor of its habitat. The tube-feet existing to the number of many hundreds in the sea-egg, are protruded, as has already been remarked, through the holes existing in each of the five double rows of perforated plates of the shell. The mechanism of their protrusion depends on the presence of a special system of vessels, known as the 'ambulacral' vessels, which carry water to the little feet, for the purpose of their inflation and distention. To this system of vessels our attention may therefore be directed.

Thus, to begin with, on the upper surface of the shell we find a single large plate, *f*, perforated with holes like the lid of a pepper-box. This plate opens into a long tube called the 'sand-canal,' *g*—a name which is decidedly a misnomer, since the function of the plate resembling the pepper-box lid is to allow water to enter this tube, but at the same time to exclude particles of sand and like matters. The sand-canal terminates in a circular vessel, *h*, which like the nerve-cord surrounds the gullet; and from this central ring a great vessel, *m*, like a main water-pipe, runs up each of the five rows of perforated plates in company with the nerve-cord. At the base of each little tube-foot is a little muscular sac or bag, and into these sacs the water admitted by the sand-canal ultimately passes. When therefore the sea-egg wishes to distend its feet for the purpose of protruding them through the shell-pores, and of thus walking by applying their sucker-like tips to fixed objects, the water in the little sacs is forced into the feet, which are thus distended. Whilst conversely, when the feet are to be withdrawn, the water is forced back, by the contraction of the feet, into the sacs, or may be allowed to escape from the perforated tips of the feet, so as to admit of a fresh supply being brought in as before.

The development of the sea-egg may be briefly glanced at by way of conclusion, along with a few points in its economic history. The animal, solid and bulky as it appears in its adult state, is developed from a small egg, which gives origin to a little body usually named the 'larva,' but which, from its resemblance in form to a painter's easel, received the name of *Pluteus*. This little body does not in the least resemble the sea-egg; it possesses a mouth and digestive system of its own, and swims freely through the sea. Sooner or later, however, a second body begins to be formed within and at the expense of this Pluteus-larva; whilst as development proceeds and ends, the sea-egg appears as the result of this secondary development, and the now useless remainder of the first-formed larva is cast off and simply perishes. Thus the development of the sea-egg forms by no means the least curious part of the animal's history, and presents a singular resemblance to the reproduction of the star-fishes and their neighbours.

The mere mention of the economic or rather gastronomic relations of the sea-eggs may appropriately form a concluding remark to the history of these animals. With our

British prejudices in favour of eating only what our forefathers were accustomed to consider wholesome, it is not likely that the sea-eggs will appeal with success to be included in the lists of our cuisine. Yet on the continent these animals are much esteemed as articles of dietary, and even of luxury. The Corsicans and Algerians eat one species, whilst the Neapolitans relish another kind; and in classic times, when variety rather than quantity or quality was the chief feature of high-class entertainments, the *Echini* were esteemed morsels at the tables of the Greeks and Romans. Here, then, is an opportunity for another Soyer to tempt the modern cultivated appetite with a new and wholesome dish. Considering that oysters and lobsters are so highly esteemed, the sea-eggs but wait a suitable introduction to become, it may be, the favourite tit-bits of future generations.

A wise philosopher—the great Newton himself—remarked concerning the limitation of our knowledge, that we were but as children, picking up at most a few stray grains of sand on the sea-shore, whilst around us lies the great region of the unknown. Our present study may not inaptly be related to Newton's comparison, since it serves to shew that even the brief and imperfect history of a stray shell picked up on the sea-beach may teem with features so curious and with problems so deep, that the furthest science may be unequal to the explanation of the one or the elucidation of the other. Whilst the subject no less powerfully pleads for the wider extension of the knowledge of this world and its living tenants—knowledge, which in every aspect, reveals things which are not only wondrously grand, but also 'fair to see.'

A GOSSIP ABOUT CRABS.

FEW members of the animal world are more familiar to all than the various kinds of crabs which the sea-side visitor meets with in plenty in his stroll along the beach. Their queer, irregular gait, their pseudo-ferocious aspect, as with claws lifted on high they menace the intruder, and their curious habits and appearance, cause them to be regarded with interest by the ordinary observer; whilst to the naturalist the crabs present very many interesting points for cogitation, not only as adult forms, but even from their youngest infancy and from the earliest periods of their existence.

As members of the great *Crustacean* class, the crabs are provided, like their familiar neighbours the lobsters, shrimps, and prawns, with a hard outside skeleton or 'shell.' This 'shell'—different in kind, it must be noted, from the shell of the oyster or mussel, &c.—is merely the outer skin of the crab, rendered hard by the deposition of limy matter; and so completely is this process of investment carried out, that we find the crab and his neighbours enclosed, even to the tips of their toes, in a hard shelly armour. Curious details have been put on record by observant naturalists regarding the periodical change of this shelly covering. For, like some veteran warrior tired of bearing his armour continually, we find the crab or lobster retiring periodically to some sequestered spot, and there lying in a quiescent state, until the shell has become loosened from its attachment beneath. The body-armour is readily slipped off; but the operation of freeing the legs from their investing shells appears to be attended with greater, and in some cases insurmountable difficulty. Occasionally, and notwithstanding all the efforts of the crab, portions of the old armour will sometimes remain firmly attached to the new suit, and thus cause not only inconvenience and pain, but absolute hindrance to the crab's progression. The old armour being safely cast aside, however, the formation of the new suit quickly takes place. Dame Nature loses no time in refitting the temporarily defenceless crustacean. Soon the new and soft skin-surface begins to secrete lime, and in a comparatively short period the crab comes forth, literally like a 'giant refreshed,' and newly equipped in a coat of mail. It may be noted that the process of casting off the hard parts extends even to the stomach of the crab and to other internal organs. Thus the horny inside layer of the stomach is periodically cast off and renewed, and the horny coverings of the eyes and portions of the gills are also subjected to this recurring alteration and change. A crab increases considerably in size after each moulting; indeed, the real object of the process is to admit of growth taking place. When clad in their armour, any increase in size is impossible of attainment; and hence the moulting of the shell is to be regarded simply as a part of the ordinary process of growth in the crustaceans. Another curious feature which has been observed in crab-existence is the process of casting off the limbs, which these forms, together with the lobsters, have been ascertained to exemplify. The sound of cannon or of thunder has, with every appearance of truth, been alleged to cause the sudden separation of one or more claws. The lost members of crustaceans are, however, capable of complete reproduction; and although occasionally a crab may be detected literally 'stumping' about with a deformed claw, yet, as a rule, the severed limb is quickly replaced by a new and perfect member.

The body of the crab corresponds, anatomically, to the head and chest firmly united together and greatly extended from side to side. If we compare the crab with the lobster, the chief difference observable between them is seen to consist in the possession by the latter of a long, jointed abdomen or tail. Hence the lobster is one of the *Macrura*, or 'long-tailed' crustaceans. But in the crab the tail or abdomen is also represented, though it must be confessed in a rudimentary condition; and if we lay the crab on his back, we shall recognise the tail in the little conical appendage tucked under the body, and to which children give the familiar name of the 'purse.' This 'purse,' bearing a few 'feet' on its undersurface, is simply an abbreviated tail; and on this account the familiar crabs are known as *Brachyura*, or 'short-tailed' crustaceans, in contradistinction to their longer-tailed neighbours the lobsters, shrimps, and prawns.

Amongst the more notable structural features which the crabs present are the stalked eyes, each of which, although apparently single, is in reality a compound organ, being composed of little spaces or 'facets,' each containing the essential parts of an organ of sight. Then we find the 'feelers' or 'antennæ' also situated in the neighbourhood of the mouth and eyes. These, in all crustaceans, number two pairs, and are composed of a series of joints supplied with nervous filaments, and perfectly adapted to subserve the sense of touch, and, as some observers maintain, probably the sense of taste also. The mouth is admirably suited for

E

biting purposes, and is provided with a series of powerful jaws by means of which the food is broken down and triturated. And, supplementary to the jaws, we find even the walls of the stomach to be provided with horny teeth, adapted for the further division of the food during its passage through the digestive system. These teeth, also seen in the stomach of the lobster, give to that organ the appearance familiar to children, and which childish imagination has named the 'lady in the lobster.'

The early life or infancy of the crab, like the development of the insect, is marked by a very distinct series of changes or 'metamorphosis.' So marked, indeed, are the changes of form which the crab undergoes in the course of development, that naturalists at first gave distinct names to the different stages, under the idea that they represented distinct and separate animals. The young crab, on leaving the egg, thus presents itself as a curious little form (fig. 16, a), pro-

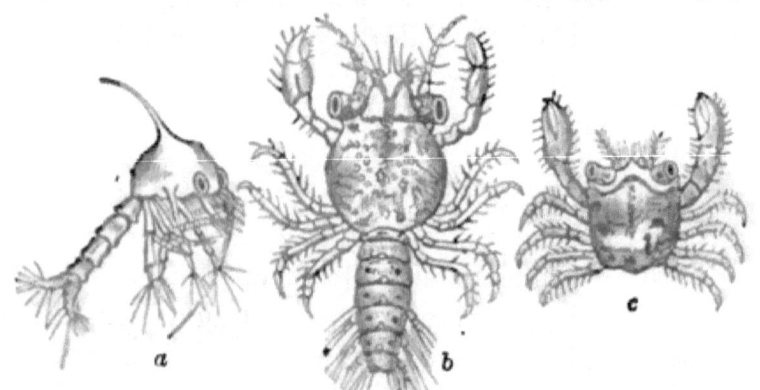

Fig. 16.—Metamorphosis of Crab:
a, Youngest stage form of crab, known as *Zoea pelagica*; b, more advanced stage of a (*Megalopa*); c, advanced stage of b.

vided with a grotesque head, somewhat of a helmet shape, which terminates behind in a long peaked process, resembling the end of a night-cap long drawn out. In front, this large head bears a pair of great lantern-like eyes, which, unlike those of the perfect crab, are destitute of stalks. And, finally, the body itself appears in the form of a slender jointed tail, by the aid of which the little crab moves in acrobatic fashion, 'head over heels.' Four pairs of appendages, representing the rudimentary limbs, are also developed in the *Zoea*, as this first stage is called; but these legs are useless as locomotive organs, and are used chiefly to draw food-particles towards the mouth. The second stage of development, known by the name *Megalopa, b*, soon succeeds the first or Zoea-stage; and the crab now makes some approach to the likeness of the perfect form. The elongated lobster-like tail is still retained by the Megalopa, and constitutes as before, the chief agent in its locomotion; but the eyes have become like those of the perfect form, and appear as stalked organs. The antennæ or 'feelers,' and the great claws, and ordinary legs are now developed; whilst the body-piece itself becomes broadened, and bears a close resemblance to that of the adult crab. The third stage, *c*, chiefly consists in the casting away of the tail, and in the assumption of the perfect form. The tail-appendage thus shrivels up and becomes of a permanently short and rudimentary nature; the body grows still broader than before; and the little creature—measuring in length only about an eighth of an inch or so—requires simply to grow in size to become recognisable as the ordinary and familiar crab. We thus observe that in its young state the crab possesses a tail resembling that of the lobster, and presents other points of affinities to its familiar neighbour; these affinities, however, in the adult state and as the process of development proceeds, being obliterated and lost.

The group of 'crabs,' popularly so called, includes very many interesting forms, which are distributed by the naturalist in different divisions of the great Crustacean class. The *Hermit* or *Soldier Crabs* are very familiar objects of our sea-coast—these creatures, each ensconced in the cast-off shell of a whelk or other mollusc, being literal hermits; whilst it can hardly be said that they exhibit hermit-like dispositions, since from their pugnacious combative instincts they as truly merit the title, 'soldier-crabs.' In marine

Fig. 17.—Hermit-crab in shell.

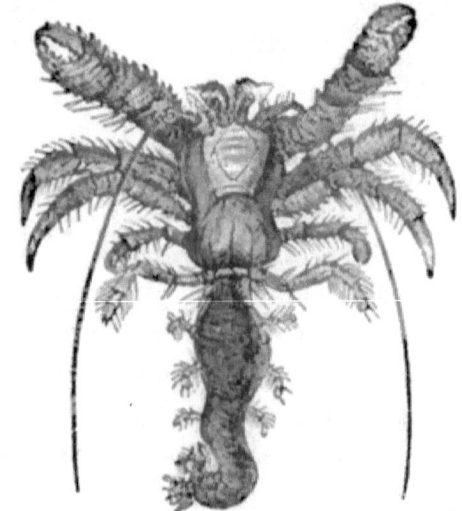

Fig. 18.—Hermit-crab: *Pagurus Bernhardus*, the Common 'Hermit' or 'Soldier' Crab removed from its shell, and shewing the soft tail.

aquaria, or in those natural aquaria, the rock-pools on the sea-beach, the hermits may afford much amusement to observers. A small crab in a large shell may frequently be seen to be engaged in combat by a larger neighbour in a small domicile; the latter frequently succeeding in ousting his lesser companion from the more roomy abode, and coolly ensconcing himself in the shell from which he has thus expelled the rightful tenant and owner. These crabs possess a soft abdomen (fig. 18), this structure leading them to seek protection in the shell; and the hinder extremity of the tail is provided with curious sucker-like feet, by means of which the animal retains a firm hold of the whorls of his habitation.

The crabs of foreign, and especially of tropical regions, exemplify several very remarkable forms. The land-crabs of Jamaica and other West Indian islands are so named from their terrestrial habits, for the pursuit of which the structure of the gills and breathing apparatus is somewhat modified from that of the ordinary sea-crabs. They inhabit burrows, which they excavate in damp or marshy situations, and appear to subsist on either animal or vegetable matter. A remarkable instinct leads the land-crabs to make an annual journey to the sea, chiefly for the purpose of depositing their eggs. This migration is said to take place during the wet season, and immense hordes of these crabs may be thus met with, marching in a straight line towards the sea. Their march is effected chiefly by night, and they are said to be exceedingly destructive to the vegetation of the districts through which they pass. The negroes trap these crabs for the sake of the flesh, which is said, when well cooked, to be tender and nutritious. The cocoa-nut crabs are so named from their habit of feeding upon these nuts. These latter are also terrestrial in habits, but appear to visit the sea more frequently than the land-crabs. The cocoa-nut crabs were supposed to climb the trees in search of their favourite fruit; but this supposition appears to be erroneous, and the more correct view is, that they feed upon the fallen nuts, which they open by first peeling off the fibrous investment, and then smashing in one end of the nut by blows from the great claws. The racing crabs of Syria present examples of forms which are able to run very quickly—these crabs being alleged to keep pace with a trotting horse.

The Molucca or king crabs, inhabiting the West Indies, the North American coast, and the Eastern Archipelago, differ materially from the ordinary crabs, in the possession of a long, spinous, sword-like tail. They are thus known to the naturalist as *Xiphosura*, or 'sword-tailed' crustaceans;

the elongated spines being used by the natives of the Eastern Archipelago to form spear-heads. The body is horse-shoe shaped, and is convex above and concave beneath. On the under surface (fig. 19, A) we find twelve feet sur-

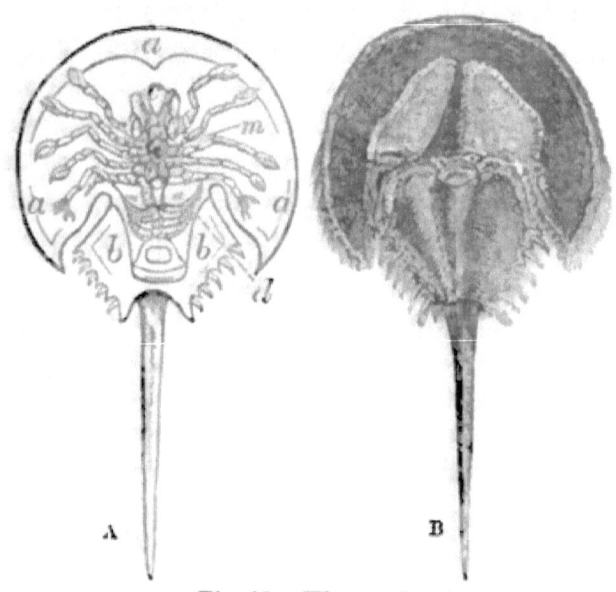

Fig. 19.—King-crabs:

A, Under-surface of 'King-crab' (*Limulus polyphemus*); *a a a*, shell; *b b*, chest; *c*, feelers; *d*, operculum, covering the branchial or breathing plates; *m*, mouth. B, upper surface of King-crab.

rounding the mouth; the first joints of these legs being armed with spines, so as to serve the purpose of jaws, and to divide the food, which consists chiefly of animal matter. The flesh and eggs of the king-crabs are highly esteemed by the Javanese as articles of diet; and hogs are said to be driven to the shore to feed upon them, the swine becoming adepts at securing their prey by turning the crabs on their backs, when, like turtles in like difficulty, they are unable to regain their proper position. The king-crabs are now kept in most of our large aquaria, where they may be seen to crawl about in a slow manner, using the long tail as a kind of lever in shifting from one position to another. One unfortunate king-crab, tenant of an aquarium-tank, had through some accident lost his tail, and was unable to hoist himself from one position into another. The attendant compassionating the helpless crustacean, fitted a wooden tail to his body, but neglected to think of the consequences to the animal which ensued on its being thus equipped; since the wooden appendage literally floated the crab out of his element, and placed the afflicted crustacean in a worse condition than before.

If some of our existing crabs might be esteemed of very large size, the geologist would convince us that the largest of our living species must be regarded as mere pigmies when compared with some of their extinct and fossil neighbours. We know of some species allied to the king-crabs, for example, which must have attained a length of six feet or more; the remains of these crab-giants being found in the rocks known to geologists as the upper Silurian and Old Red Sandstone formations. In ancient oceans, curious neighbours of our crabs, named Trilobites (fig. 20), from the three-lobed form of the body, must have crawled about in thousands, judging from the plentifulness of these fossils in the rocks, which represent the petrified bed and *débris* of these old seas. The Trilobites have completely passed out of existence; but the consideration of their immense development in the past, forms not the least interesting part of crab-history, and tends to shew us that our existing crabs have had a very long and ancient ancestry, and that they probably represent the modified descendants of large, peculiar, and often weird-like ancestors.

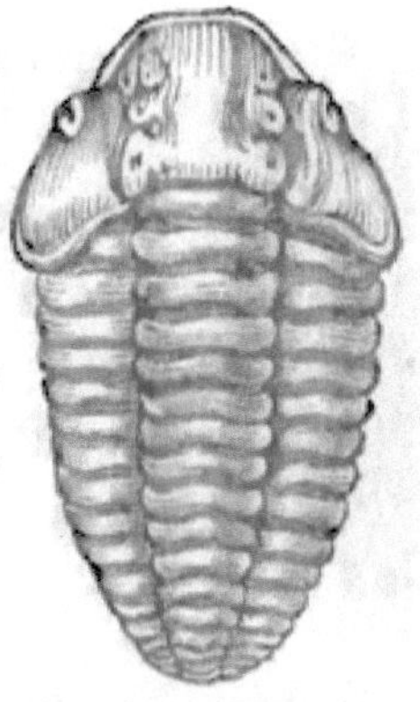

Fig. 20.—Calymene Blumenbachii.

SHELLS AND THEIR INMATES.

NO structures belonging to the animal world are better known to ordinary observers than 'shells.' It happens, however, that under this name several very different structures are included. A fishmonger, for example, asked to shew a customer his stock of 'shell-fish,' would place before the buyer such animals as lobsters and crabs, shrimps, prawns, and oysters; whilst, presuming that his stock included the greatest variety of edible molluscs, he might produce in addition the plebeian cockles, mussels, periwinkles, and limpets. A naturalist asked to classify after a scientific fashion the fishmonger's 'shell-fish,' would place the lobsters, crabs, shrimps, and prawns on one side, and the oysters, mussels, cockles, limpets, and periwinkles on the other. He would inform the fishmonger that for commercial purposes it might be convenient to designate these animals by the common name of shell-fish, but that in a zoological sense, the one set of animals and their shells would require to be widely separated from the other; inasmuch as not only are the 'shells' of the one set different from those of the other, but the included animals are also of widely diverse kind. The oyster and its neighbours are true shell-fish or *Molluscs;* the lobster and its kind are termed shell-fish only by courtesy as it were. The 'shell' of the lobster is simply a hardened skin and nothing more; the shell of the oyster being no doubt also a hard outer covering of lime, but possessing much more intimate relations with the living body it contains, than that of the crab or lobster. The shell of the latter, as we have remarked in the preceding chapter, is further periodically cast off, and as regularly replaced by a new shell, produced by the development of lime in the soft skin. Once formed, moreover, the shell of the crab and its neighbours does not increase in size, so that the growth of the body it protects has to take place during the period in which the new shell is being formed. In the oyster and its allies, on the contrary, the shell is attached by definite muscles and other structures to the animal's body, so that it cannot be periodically cast off; whilst, as we shall presently note, it is continually being added to, and exhibits a defined and regular increase in size.

The 'shell-fish' we purpose to consider, then, are the true Molluscs, of which the mussel and oyster, the snail and whelk, the cuttle-fishes, and the 'sea-butterflies,' are the typical representatives. The oyster or the mussel, as a very familiar example of a true shell-fish, may be selected for the illustration of the popular history of these animals. Whoever has tried to open an oyster, well knows the extreme difficulty experienced in gaining access to the delicate morsel within. And a very slight examination of the animal and its shell after the operation has been performed, would serve to shew that the chief obstacle which resisted the efforts of the knife, consisted of a bundle of strong fibres which connected the valves of the shell internally, roughly speaking, at about their middle. When this bundle of fibres is cut across, the oyster is rendered utterly helpless; and any one who has simply tapped a living oyster as it lay in the water with the shells slightly opened for breathing purposes, must have formed some idea of the power of these fibres to close the shell, from the quick snap with which the animal brings the shells together.

The bundle of fibres thus seen in an oyster simply represents a single great *muscle*, the function of which is to close the shell. Hence this muscle is appropriately enough named the 'adductor.' Its fibres act precisely in the same manner as that in which those composing human muscles act—namely, through the peculiar property possessed by muscle-fibre, and which is named *contractility*. In virtue of this power, muscle-fibres can shorten themselves in obedience to some stimulus, and by shortening themselves they of necessity bring together the structures between or to which they are attached. The tap on the outside of the oyster's shell represents the stimulus, which incites the contraction of the muscular fibres, thus producing the closure of the shell. Some shell-fish possess two of these muscles (fig. 25, *b b; c c*), and are thus doubly provided in the way of protective apparatus. It is clear, however, that the muscle in the oyster can play no part in the reverse action of opening the shell. In the performance of this action a most economical arrangement is witnessed. Between the shells, and at the hinge or that surface by which the shells are joined together, a stout structure may be found in the oyster and its allies, named the *ligament* of the shell. This is a highly elastic structure, and its elasticity forms the means whereby the opening of the shell is effected. For, as the ligament is greatly compressed when the shells are brought together by the muscles, it follows that when the muscles are relaxed, the elasticity and recoil of the ligament will force the shells apart. We thus note that whilst the closure of the shell is a muscular act and one involving a considerable amount of exertion, the opening of the shell is due to a purely mechanical arrangement, and is an action effected without trouble to the animal. When the shell is kept closed, the muscles are in a state of tension, since the elasticity of the ligament is opposed to the action of the muscles. But as the closure of the shell is not an act of frequent performance—the normal state of these animals being that of existing with an unclosed shell —the muscular power of the animal is thus husbanded, and the shell is kept open by the elasticity of the ligament—an action, as we have seen, requiring no effort on the part of the mollusc. No better instance, perhaps, could be found than the present, of the economical ordering of Nature's ways and works.

But supposing that our oyster has been duly opened, we may gain some ideas respecting the shell and its manner of formation from even a slight inspection of its structure. Lining the shell, and enfolding as it were all the organs of the animal, a delicate soft skin may be readily found. This skin is therefore most appropriately named the 'mantle,' and we must regard it with interest, since it is the shell-forming organ. Every true shell is formed by the mantle, and although we may sometimes find shell-like things in animals secreted by other structures, these latter are not shells strictly speaking. Thus the Paper Nautilus—that most famous of cuttle-fishes—possesses a little, fragile shell of paper-like consistence; but this latter being formed by two of the arms of the creature and not by its mantle, is not regarded by the naturalist as a true shell, and therefore does not correspond in nature to the shell of the oyster, or to that of other cuttle-fishes. Shells consist of limy matter, generally combined with a greater or less amount of horny substance. The mantle therefore possesses power of taking lime from the sea-water, and of elaborating and building up this matter in the form of the shell. The most active part of the mantle is its outer edge, or that next the margin of the shell. In this situation, it is thicker than elsewhere, and here we find the little glands on which the work of lime-formation depends. When the shell increases in size, it grows by new layers being added to its outer edge; the margin of the mantle simply forming a new layer of limy material on the part of the shell which lies next itself. And an idea of the periodical increase of shells may readily be gained by an inspection of the outer surface of many specimens in which the regular lines of growth may be perceived. The outer surface of the mantle, or that which lies

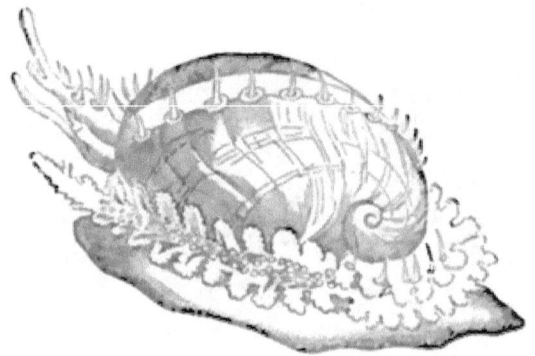

Fig. 21.—Ear-shell or Haliotis.

next the inside of the shell, forms the delicate lining of the shell, familiar to every one under the name of 'Mother-of-pearl.' The rainbow-like hues and lustres of this substance are well seen in such shells as the ear-shells (fig. 21), imported in large quantities from the Channel Islands for the purpose of affording materials for inlaid-work. The lustre of the shell-lining appears to be due to the presence of numerous fine lines or ridges which refract the light, and cause a play of colours, such indeed, as may be produced by drawing a number of fine and closely-set lines across a sheet of dull lead.

The presence of this mother-of-pearl layer in shells becomes of exceeding interest to us, when we consider that the growth of 'pearls' depends upon the quality of the shelly layer which the mantle of some molluscs is capable of forming around solid particles of matter. Pearls, in this light, are to be regarded as abnormal or unnatural formations; since they result from the presence of foreign particles, such as grains of sand accidentally introduced beneath or within the mantle-substance of the pearl-oysters and other molluscs. Around such foreign bodies, the mantle secretes the smooth pearly layer, presumably coating the intruding particle to lessen the irritation set up by its presence within the living tissues of the animal, and in due time producing what is justly considered, next to the diamond, the most famous of jewels. It is thus curious to observe that Shakspeare, in using the expression, 'if all their sand were pearl,' has unconsciously indicated the origin of the valued object.

Fig. 22.—Snail.

That the pearl-substance is actually deposited around foreign bodies is proved by the artful practice of the Chinese, who introduce substances even of comparatively large bulk into the shells of pearl-forming molluscs, with the result of procuring a coating of the pearly substance. Small metal images, thus covered, may be seen in many museums.

One of the most important structures found in molluscs is that termed the 'foot' (fig. 25, *f*). The snail (fig. 22) or whelk crawls on the broad surface formed by the foot; the organ in such a case being very appropriately named. But in other shell-fish, the foot assumes a different form and function. In the cuttle-fishes we behold it transformed into the circle of arms or feet which surround the head. In the oyster it is small and unimportant, and bears no share in the animal's life-history; the oyster being a stay-at-home and fixed creature.

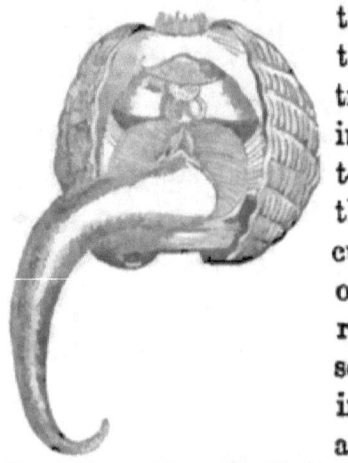

In the cockles (fig. 23), on the other hand, the foot is unquestionably the most prominent object in the body, and may be protruded to a great length from the shell; the animal using the foot as a muscular lever, and executing great leaps over the sand by its aid. The razor-shells, well known to every sea-side visitor, burrow so swiftly in the sand by means of the foot, as to defy the efforts even of the expert fisherman to intercept them. Probably the most interesting use to which the foot of the shell-fish may be put, is that of manufacturing fibres which are used for mooring the animal

Fig. 23.—Cockle, with Shells separated, shewing the Foot.

Fig. 24.—Solen, or Razor-fish (*Solen siliqua*).

to fixed objects. The 'beard' (fig. 25, *e*) of the mussel presents a familiar example of fibres formed by the foot, and whoever has tried to detach mussels from the crevices of rocks must have gained some idea of the strength of these fibres. The mussels may frequently be picked up on the sea-beach having attached to themselves stones and other objects, often of considerable size and weight; the objects being bound very firmly together by the fibres of the beard.

The manner in which the foot of the mussel manufactures the 'beard' is somewhat similar to that in which the silk-secretion of the spiders and caterpillars is produced. A special gland exists at the base of the conical pointed foot of the mussel (fig. 25, *f*)—this

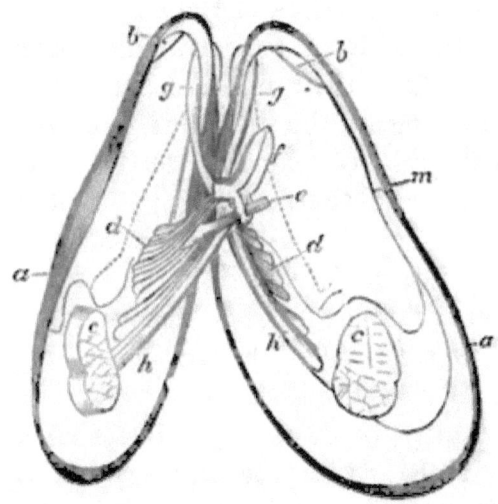

Fig. 25.—Muscles of Shell-fish (*Modiola* or Horse-mussel): *a, a*, shell margins; *b, b*, scars of anterior adductor muscle; *c, c*, posterior adductors; *d, d*, muscles attached to *byssus* or 'beard;' *e*; *f*, foot; *g, g*, anterior retractor muscles of foot; *h, h*, posterior retractors of foot; *m*, line marking the margin of the mantle.

organ somewhat resembling a ploughshare in appearance. This gland forms a fluid substance, which hardens by exposure to air and water, and becomes so tenacious that it can be drawn out in the form of a fine thread. The fluid being poured into a groove in the foot, sets in this groove as molten iron does in a mould. This first thread being pulled out of the groove by the retraction of the foot, remains attached to the organ, whilst other threads are formed in the same way so as to gradually form a bunch or 'beard.' An interesting example of the precept that 'union is strength,' is afforded by the case of the mussels which live in the crevices between the stones of a long bridge of some twenty-four arches, built across the river Torridge at Bideford in Devonshire, near the junction of that river with the Taw. The Torridge river flows with such rapidity that when the bridge was first built it could not be kept in repair, owing to the mortar being washed out of the interstices of the stones. It was noticed, however, that those stones between which mussels had accidentally lodged, kept their places despite the rapid flow of the tide; and acting on the hint thus given by nature, large quantities of mussels were brought to the bridge and placed in the crevices, with the result that the stones were prevented from slipping, and were literally tied to their places by the 'beards' of the molluscs. Special legislation renders it penal for any one to remove the mussels from the bridge, unless under certain circumstances and by express permission of the trustees.

It might hardly be deemed possible that the shell-fish could supply materials for the weaver; but the *Pinna* or Mediterranean mussel, a mollusc possessing two very thin horny shells of triangular shape, actually affords a quantity of silky fibres from the beard, sufficient to enable the Sicilians to produce a kind of soft cloth. This material is capable of being woven into gloves and stockings; and it is notable that Pope Benedict in 1754 had presented to him a pair of stockings made of the Pinna's beard-material. The Pinna's secretion is said to be mixed at Taranto, in Italy, with about a third of real silk; the material thus formed, being worked into gloves and similar articles.

The form, shape, and size of shells vary greatly. Between a size which borders on the minute, and that of the Giant Clam, which may measure from four to five feet in length, and weigh about five hundred pounds, we find many intermediate degrees; whilst peculiarities in form and shape are as plentiful as are variations in size. Of the more extraordinary forms which shells may assume, the 'Hammer Oysters' present notable examples. In these creatures we behold an accurate representation of the familiar implement just named; the borders of the shell at the hinge being extended at right angles to the body, which corresponds to the shaft of the hammer. Amongst the whelks and their class we find many peculiarities of form. The 'pelican's foot' shells are thus not unlike the objects from which they derive their name; and the 'harp shells' bear a resemblance to that musical instrument. Some are like spindles; others possess appendages of comb-like nature (fig. 26); whilst

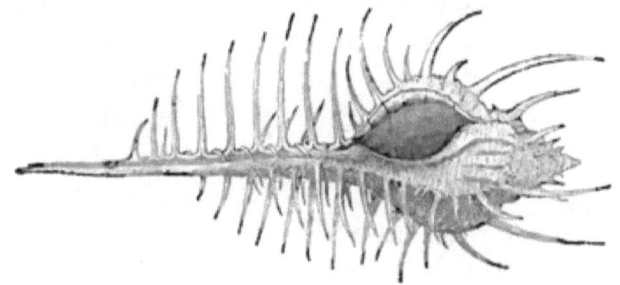

Fig. 26.—Murex tenuispina, or 'Venus' Comb.'

helmets, elephants' tusks, and other kinds of teeth, spiral staircases, and even 'caps of liberty'—the *bonnet rouge* of the French—are to be found duly represented in the cabinets of conchologists.

Little need or can be said in the present instance regarding the commercial relations of shells. The oysters and other members of the group viewed in the light of dainties, and pearl-producers, alone afford employment to thousands of persons, and give origin to an important branch of industry. Also important are those shells, such as the helmet-shells (*Cassis*) of the West Indies, from which the valued 'cameos' are cut; whilst by savage nations the little cowries are used to represent a coinage. In ancient times the purple dye —the 'blue and purple from the isles of Elishah'—which made Tyre of old so famous, was obtained from a near relative of the familiar whelk. Doubtless these molluscs were included in the products of the 'men of Tyre,' who, as mentioned in Scripture, 'brought fish and all manner of ware' to Jerusalem. That once famous city is now in ruins; but curiously enough, amidst the wreck that remains to mark its site, excavations were found in which broken whelk shells were contained, as if they had been pounded by the primitive dyers of the early times; and on the Tyrian shores at the present day the purple whelks still exist.

It might be thought that man had little to fear from the inroads of shell-fish upon his works. We have seen that the mussels contribute to insure the stability of his erections, but certain neighbours of the mussels cause dire dismay by the extent of their ravages upon human handiwork. These destructive habits, however, are not confined to the shell-fish. A little crustacean, for example, allied to the familiar wood-lice, and named the 'wood-borer' or *Limnoria terebrans*, burrows in the wood of piles, and causes much damage to piers, dockyards, and similar erections. The more harmless kinds of boring molluscs are the familiar 'borers' (*Pholas*) of our coasts, which lie ensconced in the burrows they excavate in rocks, and keep up communication with the outer world by means of the long 'siphons' or breathing-tubes which protrude from one extremity of the shell. The means whereby these shell-fish are enabled to burrow in rocks of different kinds and varying degrees of hardness, were for long undetermined by naturalists. Some naturalists broached the theory of a chemical solvent, whereby the rock was gradually dissolved away; but no such solvent could be found on examination of the animal. A second party promulgated the idea of flinty particles contained within the soft parts, as representing the

Fig. 27.—A piece of Rock bored by Pholades.

boring-apparatus; but an examination of the tissues failed to shew the presence of any such particles. The shell of the 'borers' was, however, overlooked in discussions on this subject; but when duly examined, this structure, although of delicate texture, was seen to be provided with rough file-like surfaces, which would perfectly adapt the shell for the work of excavation. That this latter view is the correct one has been proved by the inspection of these animals when at work; the shell being then seen to be moved with a rotatory motion, whilst no doubt the currents of water, which are continually passing in and out of the long breathing-tubes, materially assist in carrying off the refuse particles scooped out from the burrow. It may be remarked in passing, that the borers afford valuable evidence to the geologist of the elevation of land; since, when their burrows are found in rocks and structures now existing far above the sea-level, the geologist is led to note that the land must have risen considerably from its former site. A well-known case in point, is that of the ancient temple of Jupiter Serapis, at Puzzuoli, on the Bay of Naples, on the pillars of which the burrows of these molluscs are seen. The pillars are at the present time elevated above the sea-level, but at one time—as proved by the presence of the burrows, and by other evidence—the land must have subsided, and must have carried the pillars below the sea-level; a subsequent movement of elevation raising them to their present height.

The most famous enemy of man which the ranks of the shell-fish contain, is unquestionably the *Teredo* or ship-worm —the *calamitas navium*, 'calamity' or 'terror of ships,' as Linnæus long ago termed it. The teredo was formerly regarded as a worm; the elongated body being composed in reality by the siphons or breathing-tubes being long drawn out. The animal is, however, a true shell-fish, possessing two little shells at the front or 'head' extremity, these shells, along with certain other limy pieces, constituting the formidable boring organs of the mollusc. The ship-worm thus excavates long tubular burrows in wood, and lines the interior of the tube with a smooth layer of limy matter. All kinds of wood fail under the attack of the teredo, the hardest oak and softest pine alike ; and doubtless the softening action of the sea-water amidst which the wood is immersed, may facilitate the operations of the animal. On more than one occasion the ship-worm has caused fear and terror to reign within the hearts of a whole nation. The cost of repairing the damage done annually to our wooden piers and shipping is by no means a light charge upon the

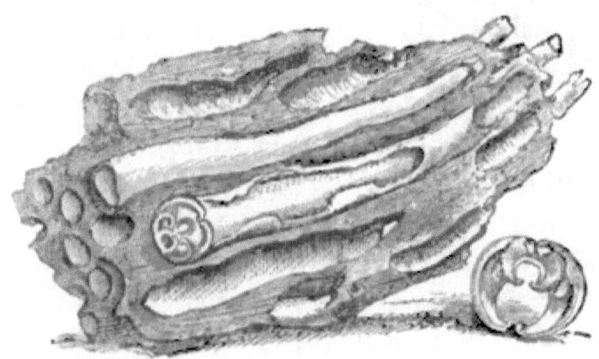

Fig. 28.—Common Ship-worm (*Teredo navalis*) in its burrows.
The detached shells are also shewn.

national exchequer; but when, as in the case of the Dutch nation, the work of the animal threatens to inundate the land, the place and power of the teredo can then be better estimated. In 1731-32, the United Provinces woke to a knowledge of the alarming fact that the ship-worm had commenced operations on the wooden piles which support the 'dykes' or banks that keep the sea from invading the land in the neighbourhood of Zeeland and Friesland. The destruction of the piles meant of course the collapse of the dykes, and the Dutch government accordingly bestirred itself, and offered large rewards for any successful plan of extermination. Lotions, paints, solutions, and chemical applications of every kind were tried without avail, and despair had well-nigh seized upon the nation. After a time, however, the teredo appeared to abandon the work of destruction it had so persistently carried on, and peace was again restored to the mind of the terrified nation.

The 'other side' to this apparently destructive action is, however, readily found in the consideration that these molluscs, by boring into and thus breaking up great masses of floating wood, serve to rid the ocean of vegetable accumulations, which, if allowed to settle in shallow waters, would render the navigation of coasts and rivers difficult or altogether impossible. The consideration of this latter aspect of the ship-worm's history, in fact, reminds us that frequently the greatest calamities and most untoward events may be shewn to be but blessings in disguise.

The shell-fish, it may be lastly mentioned, undergo a curious process of development; the young appearing at first as minute bodies, which swim freely through the water propelled by microscopic hairs or cilia. Thus the youthful oyster consists of a small body, which possesses a shell of its own, and a peculiar organ called the 'swimming pad,' by means of which the young oysters, or 'spat,' as they are collectively named, swim through the sea sometimes to immense distances from the place of their birth. Ultimately, the little roving bivalve attaches itself to some fixed object, and begins its term of sedentary life. When it escapes from the parent-shell, the young oyster is of microscopic size; at the end of one month it attains the dimensions of a pea; in six months it measures three-quarters of an inch, and a year after birth an inch and a half in length; whilst it may be said to attain its full size when it has completed its third year of life.

BUTTERFLIES OF THE SEA.

AWAY in the far north of the Arctic regions, floating in myriads upon the surface of the northern seas, and constituting vast fields of life, through which ships may sail for days and nights together, are found multitudes of small animals, to which the appropriate name of 'butterflies of the sea' has been given. To watch one of these beings pursuing its way through the waters by means of two wing-like appendages (fig. 29) springing from the sides of the neck, and to note the delicate body, enclosed in some cases in a delicate glassy shell, B, C, the comparison or resemblance to the aerial insect is by no means far-fetched or strained. In their organisation and habits, these little organisms may be found to present some points of great interest even to the unscientific reader; whilst to the naturalist they have ever afforded subjects of pleasant study and instruction.

The position of the sea-butterflies in the animal scale is of sufficiently well-determined kind. They are Molluscous animals—that is, are allied to our ordinary shell-fish, such as oysters, mussels, &c., as well as to cuttle-fishes and allied beings. Their nearest relations are undoubtedly the whelks, cowries, and other shell-fish, belonging to the great molluscan class known to naturalists as the *Gasteropoda*; and whilst some naturalists regard the sea-butterflies as forming a distinct group of themselves, others, and with every show of reason, maintain that they should be placed merely as a branch of the Gasteropod class. The scientific appellation of our sea-butterflies is the *Pteropoda*—a name signifying

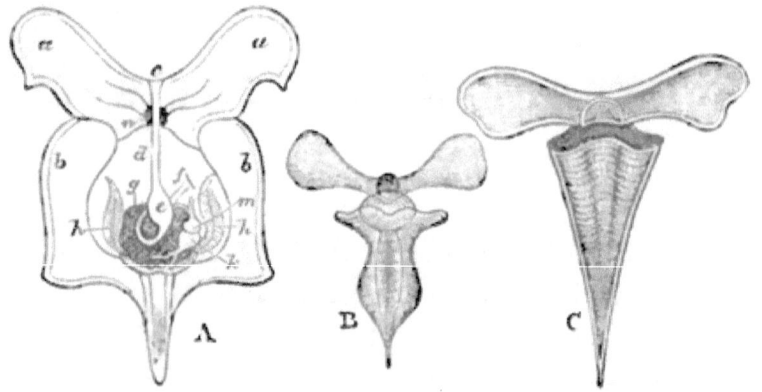

Fig. 29.—Sea-butterflies:
A, Diagram of Pteropod; *a, a*, fin-like lobes or 'wings;' *b b*, mantle or body-covering; *c*, mouth; *d*, gullet; *e*, stomach; *f*, anus; *g*, liver; *h, h*, gills; *k, m*, heart; *n*, nervous system. B, *Cleodora*, and C, *Hyalæa*, two shelled species.

'wing-footed,' and which is thus of expressive enough kind, when we consider the manner in which they flit over the watery wastes.

Besides being able to swim quickly and well by aid of their wing-like fins, the sea-butterflies can descend into the ocean-depths, or ascend from these depths to the surface, at will. They appear further to come to the surface chiefly at night or in the twilight; and as a naturalist has well remarked, each species or kind of these animals seems to have its own and special degree of darkness in which it ascends from the depths. Thus, did we know sufficient of the history of these little animals, we might be able to construct a Pteropod-clock by watching the respective hours of their appearance at the surface of the sea; just as the botanist forms a 'floral clock' by watching the times of the opening and closing of flowers.

Being 'shell-fish,' the Pteropods usually possess a shell; this latter structure, it must however be noted, not being developed in all of these beings. A very beautiful, and at the same time most representative kind of sea-butterfly, is that known by the name of the *Hyalæa* (fig. 29, C), of which kind there are several distinct species; and in this form, as well as in another well-known species called *Cleodora*, B, a shell is developed. The shell is seen to consist of a delicate glassy structure, somewhat triangular in shape, and of elongated form in Cleodora; that of Hyalæa being composed of two plates united together, and forming a small shell of globular shape. The little head-extremity of the animal, provided with its 'wings,' protrudes in each case from the front or open extremity of the shell. Another very familiar sea-butterfly is the *Clio*, which does not possess a shell, but appears as a little oblong body about an inch in length, and terminating in a lower pointed extremity. The wing-like fins, *a*, *a*, from the possession of which these animals derive their popular and scientific names, are formed by modifications of certain portions of the organ known in other Mollusca as the 'foot;' and in the present instance we find an example of the adaptation of one organ to a function widely different from that to which it is usually applied.

No part of the structure of the sea-butterflies presents more surprising details than that of the head and its appendages; the latter consisting of tentacles, jaws, and like apparatus, exercising the sense of touch and other offices. Thus, on each side of the mouth, *c*, of Clio, we discover three fleshy appendages, which at first sight might appear to consist of simple tentacles or organs of touch. When, however, we bring the microscope to bear upon these bodies, we note the interesting fact, that the surface of each is literally studded over with numerous minute specks, which, when more fully magnified, are seen to be of hollow cylindrical shape, and to contain each about twenty little suckers. These suckers may be protruded at will from their respective cylinders, so as to constitute an efficient apparatus for seizing and detaining particles of food. Thus if we consider that each of the six tentacles bears, on an average, about three thousand of the cylindrical bodies, and that each of the latter in turn contains about twenty suckers, we reach the enormous number of three hundred and sixty thousand suckers, as constituting the prehensile armament of a single Clio, itself of very small size. And imagination may assist us in its scientific aspects better than any other intellectual process, in endeavouring to form some idea of the extreme delicacy of the muscles and structures whereby the protrusion and retraction of the suckers are secured.

Two fleshy 'hoods' serve to enclose the tentacles when the latter are not in use and are retracted; and other filaments exist which may be used to subserve the sense of touch in these forms. Within the little mouth of the sea-butterflies, as also well exemplified in Clio, peculiar jaws and a curious 'tongue' exist, for the mastication of food. Each jaw is a conical structure, which literally bristles with sharp spiny teeth; and the 'tongue' is likewise studded over with recurved hooks, which also aid in rasping down or triturating the nutrient matters. And as completing the alimentary apparatus of the sea-butterflies, we find a well-developed throat, d; stomach, e; a large liver, g; salivary glands, and other addenda; whilst a heart, k, m, is also present, along with a system of blood-vessels for the conveyance of the vital fluid through the body. The breathing-organs in some of these beings are well developed, and appear in the form of delicate gills, h, h, or analogous structures, which are sometimes, as in Hyalæa, enclosed within a special chamber; but in others, such as Clio, the gills are apparently unprotected, and of indistinct nature.

A very large 'brain'—or at anyrate a mass of nervous matter, n, corresponding in function to the great nerve-centre of higher animals—is developed in the sea-butterflies, and can be discerned lying beneath the throat, and forming, in fact, a kind of internal collar around the gullet. And nerves accordingly radiate throughout the body from this central mass, and supply the various parts of the organism with feeling and vital power. Especially, as we might expect, do we find the delicate tentacles of the head to receive a large nerve supply; and we may also note the presence of two eyes, situated on the back of the neck. These latter organs are not of a very high order of development, but doubtless subserve the function of guiding their possessors in their marine flights, and are at anyrate sensitive to light.

It is very curious to observe, that, in the course of their development, the members of the higher class of the Gasteropoda already alluded to, at one period evince a strange likeness to the form of our sea-butterflies. The young whelks and their allies first appear on the stage of life as little free-swimming bodies, which move through the waters, each by means of a pair of wing-like lobes springing from the sides of the head. Observing such a form, we cannot but be struck with its close resemblance to the mature form of our sea-butterflies; a resemblance which is, however, wholly lost as the young gasteropod advances further in its development to attain its adult stage. Hence many naturalists think that the development of the Gasteropods proves these animals to be descended from ancestors or from a stage in which the Pteropod-condition was represented; the sea-butterflies, in this light, being regarded as progenitors of or co-descendants with the Gasteropods from some other and lower type of mollusca.

The food of our sea-butterflies appears to consist of the more minute marine Crustacea, which with themselves haunt the surface of the sea. Thus these small beings exist on organisms of still lesser magnitude. But in turn the sea-butterflies form a large proportion of the food of the largest of animals—the whales themselves. Drawn in myriads into the capacious mouth of the Greenland whale, with the floods of water which the great monster of the deep from time to time imbibes, the sea-butterflies remain entangled in the 'baleen' or whalebone plates of the jaws, and are thereafter swallowed as nutriment; and the species *Clio borealis*, from this latter circumstance, becomes known to us under the popular name of 'Whales' food.' Sea-birds also prey upon the butterflies of the ocean, which thus contribute largely to the support of much higher forms than themselves. In the Mediterranean Sea, on the Australian coasts, and in the Atlantic Ocean, the sea-butterflies also occur, but not in such numbers as in the far north, whither, to the very home of the Pteropods, scientific enterprise has more than once advanced on a noble mission of discovery.

Small as are all the existing representatives of the sea-butterflies, it may prove interesting to note in the last place, that, in past epochs of this world's history, several relatively gigantic members of this class appear to have been developed. In some of the oldest (Silurian) rocks, large shells of Pteropoda are discovered as fossils; one extinct species, known as *Conularia*, attaining a length of about a foot, and a breadth of fully an inch—dimensions these, of giant kind, when compared with the shells of living sea-butterflies. And in more recent rocks, the small delicate shells of our living Cleodoræ and Hyalæa may be found in a fossil state; their fossil-history proving thus to us the ancient ancestry of the 'butterflies of the sea.'

CUTTLE-FISH LORE.

OF late years cuttle-fishes have become quite established favourites in public opinion. They have been treated very much as distinguished strangers, and have accordingly been 'interviewed' to a very great extent in our great aquaria by crowds of visitors, of popular as well as scientific tastes and pursuits. Indeed, but for the establishment of these great marine menageries, together with the sensational element in Victor Hugo's romance entitled *The Toilers of the Sea*, these interesting molluscs might to this day have been reposing in their ancient quiet and retirement, which, save for the unostentatious investigations of zoologists, would have remained unbroken and undisturbed. Like Lord Byron, therefore, the cuttle-fishes have in a very short space of time awoke to find themselves famous.

The place of the cuttle-fishes in the created scale is a matter concerning which much misconception appears to exist in the popular mind. We ourselves have heard a company of intelligent visitors gravely debating in front of the cuttle-fish tank in the Brighton Aquarium, regarding the nature of its tenants; and so varied and erroneous were the opinions expressed, that the recital formed a strong argument in favour of the necessity for the cultivation of natural science in our schools and by the people at large. The cuttle-fishes are very nearly related to our familiar oysters, cockles, whelks, and other molluscs. If a very homely comparison be permissible, we might appropriately designate any cuttle-fish as a kind of cousin to the oysters and their allies. Both animals are 'shell-fish' in the true acceptation of the term—only that the cuttle-fish in the great majority of cases has its shell enclosed *within* its body, instead of being an external and protective structure—and the bodies of both animals, with those of all other molluscs, are built up on one great plan of structure. The modifications of this plan which are seen in the cuttle-fishes consist in

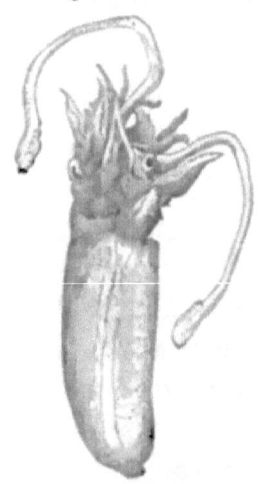

Fig. 30.
Common Squid or *Loligo*, a ten-armed cuttle-fish.

the special development of the head, which bears curious arms, feet, or tentacles, provided, in all cuttle-fishes except one, with suckers; whilst the body is enclosed in a soft muscular skin named the 'mantle,' this latter structure containing the gills, which number two or four. Another feature of the cuttle-fishes consists in the presence of a tube or 'funnel' opening just below the head, on the front of the body; this funnel being used for the forcible expulsion of the water which has been used in breathing. From the possession of numerous arms or feet surrounding the head, the class to which the cuttle-fishes belong has been named *Cephalopoda*, or 'head-footed' mollusca. Linnæus first applied this name to these creatures, and the term is still retained as a most appropriate and suggestive title for the cuttle-fish group.

Not the least curious part of cuttle-fish history is that connected with the structure of the suckers with which the arms are provided. Each sucker consists of a firm ring of cartilaginous material (fig. 31, *a*), enclosing a shallow cup-like space, the floor of which is composed of muscular fibres. This sucking-disc is perforated by a central aperture, *b*, within which a muscular plug or piston, *c*, capable of being protruded or retracted, is contained. Such is the essential conformation of the suckers, and once grasping this general idea of their structure, we shall find no difficulty in understanding the mode in which these organs are brought into play. For the adhesion of the suckers is effected in virtue of a well-known natural law, familiarly illustrated in the action of the leather sucker, by means of which the school-boy lifts large stones and other bodies of great or considerable weight. In using the sucker, the boy presses the flat disc of leather on the stone, and kneads it until he has exhausted the air below the disc. In other words, he produces a vacuum or space unoccupied by air, and the sucker is made to adhere firmly to the stone by the pressure of the superincumbent air or atmosphere which, as every one knows, presses upon each square inch of surface with a pressure of about sixteen pounds. Similarly, then, to the mode of operation of the school-boy's sucker, is the adhesion of the suckers of the cuttle-fish secured; only, in the latter case, the *modus operandi* is after a much more elegant and readier

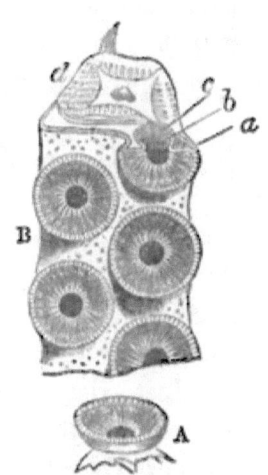

Fig. 31.—Suckers of Cuttle-fish:

A, a single sucker, side view. B, a portion of one of the tentacula, with several suckers, front view; *a*, cartilaginous ring; *b*, central cavity; *c*, piston; *d*, section of the tentacle.

fashion than in the school-boy's simple experiment. The cuttle-fish has merely to apply the suckers, then to withdraw or retract the little muscular pistons, when a vacuum is at once produced in each disc, and the perfect adhesion of the suckers secured. Whilst, in order to release its grasp, the animal has merely to push up the pistons, when the vacuum is destroyed by the admission of air below the suckers, and the arms are thus once more set free. When we consider the number of the suckers—averaging, in the commoner forms, from one hundred and fifty to two hundred suckers or more on each arm—with which each tentacle is provided, and reflect on the rapidity with which each sucker can thus instantaneously be brought into play, we must admit that the grasp of the cuttle-fish is of the most powerful and efficient kind. Nor is the simple structure just described the only apparatus by which the adhesion of the suckers is procured. In some species represented by the 'Hook Squids,' we find the margins or edges of the suckers prolonged into horny hooks, which, like the claws of a cat, can, in certain cases, be protruded or retracted at will.

The cuttle-fishes move about in a very active fashion, as any one who has attentively watched these animals in an aquarium may testify. They crawl head downwards by means of the feet and their suckers, passing from rock to rock with the agility of acrobats. They can also propel themselves backwards (fig. 32) by means of the jets of water emitted from the 'funnel' already mentioned. This mode of locomotion exemplifies an exceedingly beautiful provision for making use of an otherwise useless commodity. In other words, the water thus used for the purpose of locomotion consists of effete water, or that which has been used in the gills for purifying the blood of the creature, and which, instead of being simply ejected from the body, is thus economised and made subservient to the movement and locomotion of these forms. The assertion of the wise man, that 'there is nothing new under the sun,' finds in this hydraulic apparatus of the cuttle-fish an apt illustration; since experiments in the propulsion of ships by water-power, cannot but at once and forcibly recall to mind the analogous means of movement in these forms.

Fig. 32.—Cuttle-fish (*Sepia officinalis*):
The upper figure represents the animal in the act of swimming backwards; the lower figure, the animal at rest. (From specimens at the Crystal Palace Aquarium in 1876.)

Nothing, at anyrate, can be more elegant than the movements of these creatures in this latter mode; and the author remembers idling away—in a not unprofitable manner—a whole summer afternoon, watching a number of common cuttle-fishes or Squids propelling themselves over the surface of a calm sea in the most graceful manner possible. The tip of the body, like the bow of a ship, protruded just above the surface of the water; whilst below, the head and

G

funnel were at work, and the triangular fin terminating the body also performed its part in serving as a front rudder to guide and direct the way. The entire process reminded one of a swift vessel, propelled by efficient apparatus and guided by a skilful crew.

The possession of a bag or sac, secreting an inky fluid, is well known to form a characteristic feature of the cuttle-fishes; whilst the metaphor describing a voluminous writer as a kind of human cuttle-fish or 'ink-squirter,' is by no means inappropriate, whatever opinion may be expressed regarding its politeness. The cuttle-fish, when hard pressed by its enemies, emits this inky fluid from the funnel; and escapes under a literal cloak of darkness, produced by the rapid diffusion of the ink through the surrounding water. This fact may have forcibly suggested to many of the classical poets the idea of causing their heroes to escape under the shadow of a protecting cloud sent by some kind deity. And Oppian well describes the habit of the cuttle-fish in his lines:

> A pitchy ink peculiar glands supply,
> Whose shades the sharpest beam of light defy.
> Pursued, he bids the sable fountains flow,
> And, wrapt in clouds, eludes th' impending foe.
> The fish retreats unseen, while self-born night
> With pious shade befriends her parent's flight.

The best known kinds of cuttle-fishes, and those which have been kept in our aquaria, are the genera Sepia, Loligo, Sepiola, Octopus, and Eledone. Of the first genus, the Mediterranean Sepia (fig. 32), sometimes found on our own coasts, is a very familiar species. The Loligos include the well-known Squids (fig. 30); and the Sepiolæ are nearly allied to the Sepia itself. The Octopi are the most famous of aquarium-inhabitants, and differ from the three preceding kinds in possessing eight arms only; the Sepiæ and Squids possessing ten arms, of which two are longer than the others (fig. 30), and are provided with suckers at their extremities only. The Eledone is also an eight-armed cuttle-fish; and a very familiar species of this genus is the *Eledone moschata,* so named from the musk-like odour it emits.

A peculiar, and at the same time rare—as far as regards the *living* animal—species of cuttle is the little Spirula (fig. 33), the shells of which, named 'post-horns,' from their resemblance to the French-horn of the old coaching-days, are cast up in thousands on the shores of New Zealand. The actual animal, curiously enough, is rarely met with. The *Challenger* expedition obtained a specimen of the spirula shell and its contained animal, in the dredge off the Moluccas. This specimen makes the fourth perfect specimen which has been procured. It came up from a depth of 360 fathoms, and strange to say, appeared to have been swallowed by some deep-sea fish, and to have been vomited into the dredge as its captor was drawn upwards to the light of day.

Fig. 33.—Spirula.

By far the most celebrated of the cuttle-fishes is the *Argonauta Argo* of the naturalist—the Paper Nautilus of the everyday reader—so called from the delicate nature of its shell, which, by the way, is not a 'shell' in the true sense of the term. This structure is secreted by two of the arms, which are broad and flattened; and hence it does not correspond to the true 'shells' which we have already seen to be secreted by the 'mantle.' No animal has been more celebrated in fable, romance, or poet's song, than the paper nautilus. The ancient bards, ever quick to trace analogies and resemblances, gave it the credit of first suggesting to adventurous man the idea of a ship and of sailing on the sea. Oppian thus says that the

> Ship-like fish the future seaman taught.
> Then mortals tried the shelving hull to slope,
> To raise the mast, and twist the stronger rope.

Nor was this idea altogether inconsistent, when we consider the thoughts which engendered it. We thus find Aristotle and Pliny telling us that it floats on the surface of the sea, using its six ordinary arms as oars, and the two expanded arms as sails. But these ideas have not been confined to

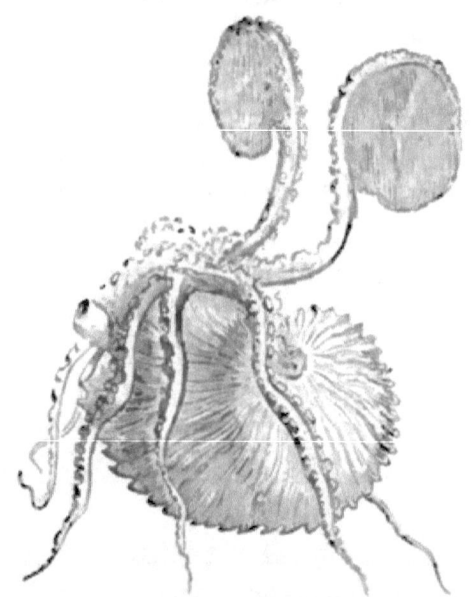

Fig. 34.—The Paper Nautilus.
The animal is represented with the two expanded and shell-forming arms detached from the shell.

the authors of olden times; the poets of modern days have, in many cases, but repeated the similes of their predecessors. Byron, for example, speaks of

> The tender nautilus who steers his prow,
> The sea-born sailor of his shell canoe,
> The ocean Mab, the fairy of the sea.

Montgomery, the poet of the sea, describes the paper nautilus in the following terms:

> Light as a flake of foam upon the wind,
> Keel upward from the deep, emerged a shell
> Shaped like the moon ere half her horn is filled;
> Fraught with young life, it righted as it rose,
> And moved at will along the yielding water.
> The native pilot of this little bark
> Put out a tier of oars on either side,
> Spread to the wafting breeze a twofold sail,
> And mounted up and glided down the billow
> In happy freedom, pleased to feel the air,
> And wander in the luxury of light.

Or, lastly, we find Pope in his *Essay on Man*, advising us to

> Learn of the little nautilus to sail,
> Spread the thin oar, and catch the driving gale.

These quotations, then, will serve to prove the existence of the prevalent idea that the argonaut is literally a living ship and sailor. However beautiful the simile may be, science unfortunately cannot maintain its truth or correctness; and in spite of poetical fancies, it must be declared that the argonaut never uses its webbed arms as sails, or its ordinary arms as oars; and that it never floats thus on the surface of the sea, but crawls head downwards like any other and mundane cuttle-fish over the bottom of the ocean, or propels itself backwards in the water by means of its funnel and jet of water. The expanded sail-like arms of the paper sailor are in reality used to attach the shell to the body, and not only repair the shell when injured, but secrete and form it at first.

The Pearly Nautilus forms the only remaining member of the group of which mention may be made in the present instance. The nautilus deserves notice principally because it links itself in a very striking manner with the past history of the cuttle-fishes. Its shell, unlike that of the paper nautilus, is a true shell, in every sense of the term. The pearly nautilus stands by itself in this particular, and in almost every other respect also. It exists as the sole survivor of a race of cuttle-fishes now only found as fossils in the rock-systems of the globe. It possesses four gills, whilst

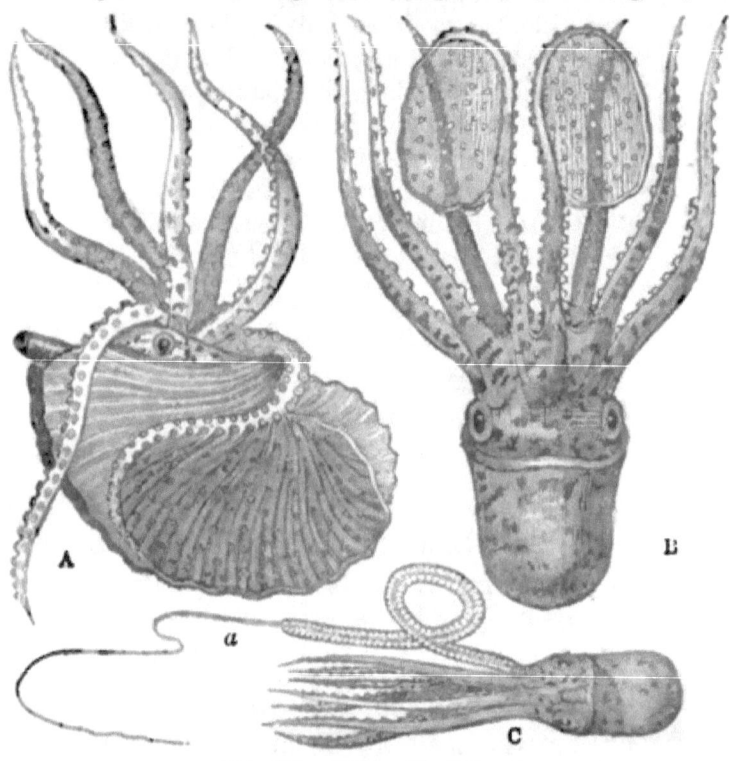

Fig. 35.—Paper Nautilus:

A, Paper Nautilus (*Argonauta Argo*), shewing the funnel, and the expanded arms clasping the shell. B, the *Argonaut* removed from the shell, and shewing the expanded arms. C, Male Argonaut destitute of a shell.

all other living cuttle-fishes possess but two. It possesses numerous arms destitute of suckers; all other living cuttle-fishes possessing at most ten arms, and these are provided with suckers. It has no ink-sac, whilst all other living members of the group possess this gland; and lastly, its shell is divided into a number of chambers, and is of highly characteristic form and structure. Like 'the last of the

Mohicans,' the nautilus may be said to live in the past; and if the ideas of humanity are at all applicable to cuttle-fishes, one might be disposed to think that its very singularity might tempt it to wish itself numbered with its

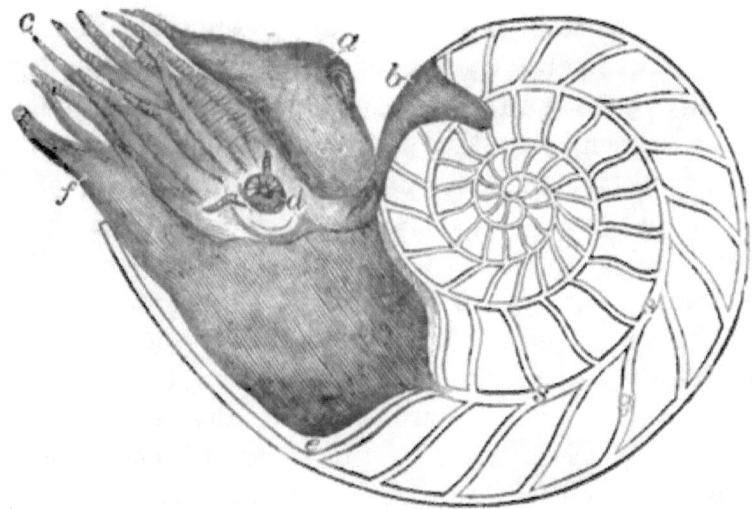

Fig. 30.—Pearly Nautilus:
The 'Pearly Nautilus' (*Nautilus Pompilius*), with the shell in section, and shewing, *a*, upper fold of mantle or 'hood;' *b*, dorsal fold of mantle; *c*, tentacles; *d*, eye; *f*, funnel; *g*, one of the partitions of the shell; *s, s*, siphuncle or tube piercing the shell-partitions.

extinct and fossil neighbours. Its chambered shell affords us a clue to its mode of growth; since, in the early life of the animal, the shell consisted of but a single compartment. As its body grew and increased in size, its habitation became too small. It consequently set about adding to its house, and formed a second and larger chamber, shutting off the first apartment from the newer chamber by a partition-wall. Similarly, when the second chamber became too contracted, and when by the exigencies of its growth it was forced to leave it, the animal constructed a third and still larger compartment, and in like manner built a partition-wall between the second and third chambers. In this manner it formed the chambered shell, seen in section in the illustration. The chambers are not wholly shut off from each other, communication between the various apartments of this curious house being kept up by means of a tube or 'siphuncle' (fig. 36, *s, s*), running through the partitions between the chambers, and terminating in front in the body of the animal. The function of this tube may be that of maintaining a certain degree of vitality in the shell; whilst the entire structure of chambered apartments doubtless serves to enable the animal to rise or sink in the water at will. The chambers of the shell, in all probability, can be filled with some gas or water; this arrangement, by altering the specific gravity of the animal—that is, by altering its weight as compared with the weight of the surrounding water—enabling it to approach the surface, or, on the other hand, to sink to the depths of the ocean. Locomotion in the pearly nautilus is effected by means of the ejection of water from the funnel, in a manner similar to that described in the case of its more commonplace neighbours.

The economic history of the cuttle-fishes shews us that these forms are highly esteemed in certain quarters of the globe as articles of diet. With our native proclivities in favour of solid and accustomed fare, we may be somewhat disposed to regard the cuttle-fish with little favour as an element of daily or even occasional consumption, but if trustworthy accounts are to be at all relied upon, we have yet to discover both a dainty and wholesome dish in cooked cuttle-fish. Among the natives of Southern Europe, the cuttle-fish is in high repute as a domestic viand. The ordinary *Sepia* of the Mediterranean may be seen regularly exposed for sale in the market-place of Naples, and many other Italian coast towns. And the Neapolitan housewives have a sure test of the freshness of their fish in the brilliancy of its colours; since, when decomposition sets in, the delicate hues vanish, and the dullness of the colour proclaims the unfitness of the fish to be used for food. The Chinese, too, who, by the way, are not at all over nice or particular in the matter of diet, esteem the cuttle-fish as a high luxury; and the ancient Greeks and Romans appear to have accounted them in a similar manner. We are told, for example, that at a certain classical marriage-feast a dish of a hundred cuttle-fishes formed one of the chief items in the bill of fare. Indeed, one gourmand of old, Philoxenus, a poet of Cythera, was so fond of the cuttle-fish or *Polypus* ('many-feet'), as it was named by the classics, that he well-nigh killed himself by a surfeit of his favourite viand. The occurrence is well related in the following lines :

>   Of all fish-eaters,
> None, sure, excelled the lyric bard Philoxenus.
> 'Twas a prodigious twist!   At Syracuse,
> Fate threw him on the fish called ' Many-feet.'
> He purchased it and drest it; and the whole
> (Bate me the head) formed but a single swallow.
> A crudity ensued; the doctor came,
> And the first glance informed him things went wrong.
> And 'Friend,' quoth he, 'if thou hast aught to set
> In order, to it straight; pass but seven hours,
> And thou and life must take a long farewell.'
> ' I 've nought to do,' replied the bard; 'all 's right
> And tight about me.   . . . .
>   I were loth, howe'er,
> To troop with less than all my gear about me;
> Good doctor, be my helper, then, to what
> Remains of that same blessed " Many-feet ! " '

Another classic poet mentions the most approved modes of cooking the cuttle-fish, and his ideas, freely translated, may be thus rendered :

>   Good-sized polypus in season,
>   Should be boiled—to roast them 's treason ;
>   But if early, and not big,
>   Roast them; boiled, arn't worth a fig.

The investigation of the domestic life and economy of the cuttle-fishes reveals some interesting and anomalous features as exhibited in aquaria. We some time ago received some interesting particulars regarding the pranks of some octopi which lived in Dr Dohrn's famous aquarium at Naples. Three octopi shared a large tank with three lobsters; the six being original proprietors and tenants of the miniature sea. Any new-comer, however nearly related he might be to either the crustacean or cuttle-fish tenants, was invariably received with demonstrations of the most hostile nature. A lobster and octopus battle is certainly a novelty in the way of animal combats, but such a fight actually occurred in the Naples arena. A lobster-giant, who had previously exhibited his prowess in crushing with his great pincher-claws the skull of a turtle as easily as if the reptile's head had been a nut, was introduced into the happy family circle in the octopus tank. Immediately, the largest octopus gave battle to the crustacean; the lobster, early in the fight, seizing one of the soft, pliant arms of his opponent in his claws; the octopus managing, however, after a time, to withdraw the captured member. Day by day the combat dragged out its weary length, sometimes one side being temporarily victorious —as when the lobster lost a large claw—and sometimes the other. At last the combatants were separated, the lobster being placed in a new and unappropriated domain in an adjoining tank.

Now comes the strangest part of the history; for the octopus, as if seized with the passion which, if exhibited in humanity, we should term one of 'dire revenge,' climbed over the partition separating the tanks in search of his enemy, and, having found him, proceeded to wage war anew. The result was most disastrous to the crustacean, for the octopus was found, we are told, with the lobster in his clutches, literally torn into halves. Thus, to natural ferocity, we find the octopus unites immense agility and a stolid persistence. This same cuttle-fish extended no sympathy to his own species; for when two others—in addition to the two who had from the first been his companions—were introduced into his tank, he chased them from the water, and forced them to take refuge on the dry rocks above. Another octopus, in a British aquarium, pulled out the plug of his tank, and brought death on himself and all his companions in a single night.

Some interesting observations have been recorded regarding the propagation of these creatures. Like other Molluscan forms, they reproduce their species by means of eggs. Some cuttle-fishes, such as the Squids and Sepiæ, do not mount guard over their eggs, which, like those of some molluscs, are enclosed within tough capsules, and are aggregated into masses known to sea-side visitors by the name of 'sea-grapes,' and the like. The octopi, on the contrary, make a nest of large stones, and therein enclose and guard their eggs—which, however, are not encased in capsules—with jealous care. The female is thus exceedingly careful of her progeny, and continually pours upon the eggs currents of water from the 'funnel;' the presumed object of this latter procedure being that of keeping the water around them in a duly aerated state. The male octopus appears, unfortunately, to be afflicted with cannibal-like propensities; for we grieve to learn that one of the chief cares of the anxious mother is to ward off the attacks of her unfeeling spouse, in his endeavours to make a repast of the developing progeny.

As may be imagined, the cuttle-fishes are very well provided in the way of organs of sense and perceptions. Their large prominent eyes are well adapted for acute vision amid their dull watery abodes; and no less acute are their hearing powers, for, although destitute of outer ears, they yet possess well-developed internal organs of hearing. The sense of touch may be subserved by the muscular arms, which may be made, as we know, to move with great rapidity, and in all conceivable directions. And as to the 'emotions' of our cuttles, strange as it may be thought, their feelings of anger and of pleasure may be tolerably easily guessed at by watching, as in the human subject, the play of colours on their skin. Not only does our cuttle-fish blush, but he may literally blush almost any hue he pleases; this property of changing colour residing in the little colour-cells that lie beneath the delicate transparent outer skin. By altering the position of these cells, various and rapidly-changing hues of colour are produced; and the famous kaleidoscopic power of the chameleon is, in truth, thrown completely into the shade by the talents, in this respect, of the lower cuttle-fishes.

The development of perceptions and intelligence in the cuttle-fishes attains a higher stage than might at first sight be supposed. The 'brain-mass' is large, and is enclosed in a kind of skull, formed by a gristly box. Whether or not this brain may 'think' is a very difficult question to determine; but we learn that in time the cuttles come to know and to recognise their keepers in aquaria, and may even exhibit symptoms of pleasure on the approach of friends. Fishes, and even those stolid emotionless reptiles the tortoises, come in time to know the hand that feeds them. And one can scarcely feel surprised that the active, wary cuttle-fishes should care to know the providers of their daily wants; whilst the circumstance enlarges the application of Schiller's aphorism, that hunger is one of the powers that rule the world.

In that certain members of the cuttle-fish group may become immensely developed in size over their ordinary neighbours, and appear as veritable giants of their race, these animals present another point of remarkable interest. Stories of gigantic cuttle-fishes, as every classical scholar knows, are frequently to be met with in the records of the ancient naturalists. Indeed, there are very few maritime countries in which, under some legendary name or other, the histories of giant members of this group may not be met with. The 'Polypus' of the ancients, the 'Kraken' of the Scandinavians, the 'Picuvre' of the Channel Isles, and the 'Devilfish' of Victor Hugo, are all so many names expressive of a belief in giant cuttles, the size and powers of which, as may be guessed, are never understated in any of the stories or legends referred to. Until very recent times, naturalists were inclined to be sceptical regarding the truth or probability of such stories. Stray examples undoubtedly had occurred now and then, of fragmentary portions, of what must have been at anyrate exceptionally large cuttles, being found. But it was often difficult, or even quite impossible, to separate the grains of truth which might be contained in any tale, from the copious husks and chaff with which the fertility of human imagination, together with the natural lapse of time and frequent repetition, invariably tend to surround the original germ or incident. Thus one may naturally be disposed to allow much for exaggeration, when we find Pliny of old telling us of giant cuttles; or when the mediæval De Montfort and Olaüs Magnus write of the Northern Kraken, of which latter creature, the worthy but credulous Bishop Pontoppidan remarked, that it was 'liker an island than a beast.'

In Captain Cook's first voyage, however, the remains of a giant cuttle were found; and parts of this specimen, preserved in the Hunterian Collection of the London College of Surgeons, shew that its total length must have exceeded six feet. A very large specimen was met with in 1861 between Teneriffe and Madeira by the French corvette *Alecton*, the length of this specimen inclusive of the arms being estimated at about fifty feet. But of late years the carcases of several very large cuttle-fishes have been met with off the coasts of Newfoundland; and a photograph representing the head and tentacles of one of these specimens now lies before us. The tentacles or arms were ten in number in this case. Two were elongated, as in the Squids, these measuring twenty-four feet in length. The eight shorter arms each measured six feet in length, and ten inches in circumference at their bases, where they joined the head. The ten arms were provided with about eleven hundred suckers; and the eyes measured each four inches across. The length of this specimen—inclusive of the arms—was thirty-two or thirty-three feet.

The question which at once presents itself to naturalists for consideration on inspecting such a form, is, whether or not it constitutes a new species, or is merely a giant member of an already known one? The cautious naturalist would hesitate to construct a new species without satisfactory evidence of more marked deviation from any type than that afforded by mere size—always a delusive test with regard to animal structures. But the mere occurrence of these monster cuttle-fishes is in any case of exceeding interest, as opening up a new field for discovery, which may in time tend to throw light even on the great 'sea-serpent' mystery itself. Why, if giant cuttles really exist, may not giant snakes also exist—largely developed individuals of the numerous species of sea-snakes known to exist in warm seas? Zoology, at least, offers no objection to the latter suggestion. And after all, in cuttle-fish history, as in most other things, the truth is certainly stranger than fiction.

We have thus considered the present history of these forms; and it only remains in the briefest possible manner to allude to the history of their past. It is needless to remark how closely and inseparably the past history of living beings is bound up with their present; and how, conversely, the yesterday of these and all other forms, must and only can be read by the aid of our knowledge of them as they exist to-day. Geologically speaking, then, the past of the cuttle-fishes has been, as far as the animals themselves are concerned, exactly the opposite of their present condition. In the seas of by-gone periods in our earth's history, cuttle-fishes swarmed in abundance, it is true, but these old representatives of the group differed widely from their modern

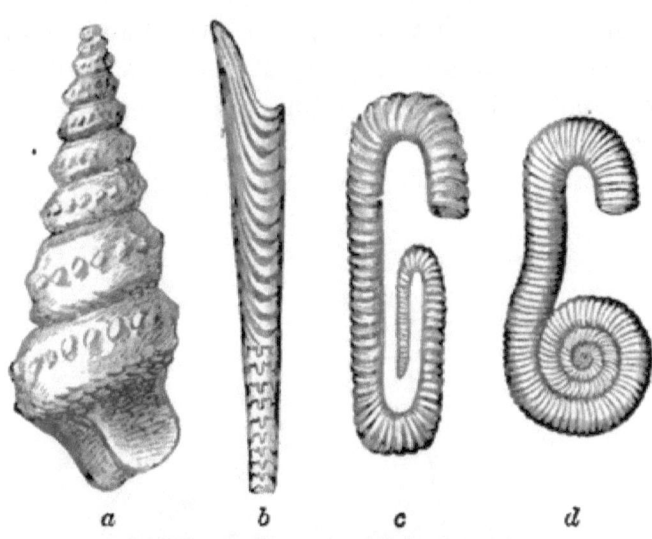

Fig. 37.—Shells of Fossil Cuttle-fishes:
a, Turrilites; *b*, Baculites; *c*, Hamites; *d*, Scaphites.

relations. The cuttle-fishes of the past, and more especially of the earlier seas of our world, possessed four gills; and associated with this particular structure, they possessed external shells, in many cases of beautiful symmetry and curious form (fig. 37). They resembled in essential structure the solitary Pearly Nautilus of our own day—the only four-gilled cuttle-fish that has remained to represent the many families of the past. Conversely, in the present day, the two-gilled cuttle-fishes attain the maximum of their development; though in future ages it may be that these forms will repeat the history of their four-gilled neighbours, and be succeeded by a new and still stranger race of forms. The two-gilled cuttle-fishes, then, as compared with the four-gilled forms, are a very much younger branch of the class. The former are totally unrepresented in the oldest of the life periods into which the geologist divides time past; whilst, on the other hand, the four-gilled forms make their appearance in some of the oldest rock-formations in which the remains of living things have been preserved. Strange it is to think that in a few years more, the Pearly Nautilus itself will probably repeat the history of its fossil relations, and will also be numbered among the forms of the past.

ON SOME ODD FISHES AND THEIR COMMONPLACE NEIGHBOURS.

THERE is no word more widely applied or more carelessly used in the English language than the word 'fish.' From being employed to denote peculiarities in man himself, under the designation 'queer fish,' to designating beings so low in the scale of creation as 'star-fishes' and the like, the word has had attached to it very varied and equally diverse meanings. As naturalists, we certainly should limit the term to denote the lowest group of that great sub-kingdom of animals to which man himself belongs. This division every one must know as that of the *Vertebrata*, a word which may be used in a popular sense, as corresponding to the expression 'backboned' animals. At the head of this group man and quadrupeds are found, whilst the fishes form the lowest class in the division.

There are few groups of the animal world more interesting to the ordinary observer than that of the fishes. To survey the various forms and shapes presented by these animals as displayed in a great museum, should prove a sufficient incentive to gain a more intimate acquaintance with the class; and when, even in a popular sense, we investigate the structure and habits of fishes, the study increases in its fascination and interest. Whilst if we reflect that on a knowledge of the habits of fishes, of their distribution in our oceans and seas, and of the special products which many of them offer for our use and luxury, the commercial success of our fisheries depends, it can need no further argument to convince us that, after all, there is something of great practical benefit to be derived from the study of zoological science.

It is not our intention at present to say anything regarding the commercial or economic aspects of fishes, and even their general habits must be very briefly touched upon. We rather aim at giving some account of the structure of the fishes, and at noting such peculiarities in their habits and life as may prove most interesting to our readers. Primarily, then, we find that fishes may be recognised by having the body usually, but not always, covered with *scales*, of various forms and kinds. Then, secondly, we have the limbs represented by certain *fins*; and, thirdly, we find almost all fishes to breathe by *gills* during the whole of life. These three points are, in the main, sufficient to distinguish fishes from their higher as well as their lower neighbours. The scales which cover the bodies of fishes present great diversities in shape, size, and appearance. Some fishes thus exhibit an utter want of scales; whilst others, like knights of old, are encased in a veritable suit of scaly armour. The lampreys, and their curious neighbours the hag-fishes, are destitute of scales; and in our familiar eels, the scales are very small and insignificant. Such fishes, however, are amply compensated for the want of scales by the power they possess of throwing out from the skin a vast quantity of glutinous or oily matter, technically named *mucus*. The presence of this secretion, which has given origin to the phrase 'as slippery as an eel,' serves to protect the surface of the body, and no doubt also assists in the easy progress of these fishes through the water. So large is the quantity of this oily matter which the Hag-fishes (fig. 45) can emit from their body, that one form has received the specific name of *glutinosa;* the fish being able in this way to literally convert the water of the vessel in which it is contained into a jelly-like mass. The familiar blennies, found in rock-pools after the tide has receded, are also able to emit a large amount of this glutinous fluid.

Illustrating the opposite extreme of the development of scales, we find such fishes as the Bony Pikes of North American lakes and rivers, the bodies of which are covered with an armour of closely fitting and overlapping scales or

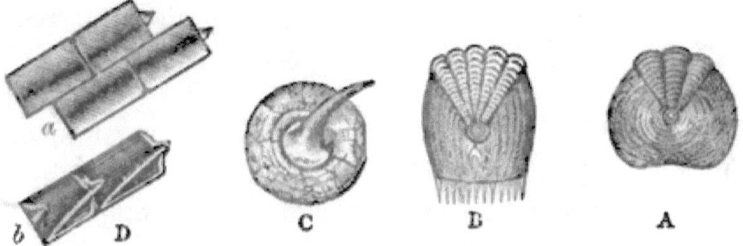

Fig. 38.—Scales of Fishes:

A, 'Cycloid' or circular scale (Salmon). B, 'Ctenoid' or comb-like scale (Perch). C, 'Placoid' scale (Ray). D, 'Ganoid' scales from a fossil fish (*Amblypterus striatus*) (Carboniferous): *a,* upper surface; *b,* under surface, shewing articulating processes.

plates, named *ganoid* (fig. 38, D), from their shining appearance (Greek *ganos,* splendour). The scales of this fish are said to be employed in the manufacture of the little 'mother-of-pearl' buttons, so commonly used. Many fossil fishes (fig. 39) were also abundantly provided with these hard bony plates; and our living sturgeons (fig. 40) possess scales of similar nature, although in the latter fishes they do not completely cover the body. The bright silvery scales of the herring and its neighbours are thin structures (fig. 38, A), and are very easily detached from the skin;

and a curious form of scale is seen in the perches (fig. 38, B); the hinder edge of each scale in the latter case being cut into comb-like teeth. In the sharks, skates, and rays, the scales are small and horny, and are often provided with little spines (fig. 38, C). If we draw our hand along the back of a dog-fish from tail to head, as when we stroke a cat's back the wrong way, we feel numerous small projecting points, borne on the scales. The rough skin surface thus produced is frequently used under the name of 'shagreen' in the manufacture of spectacle-cases and like articles, and is also employed for polishing the surface of wood.

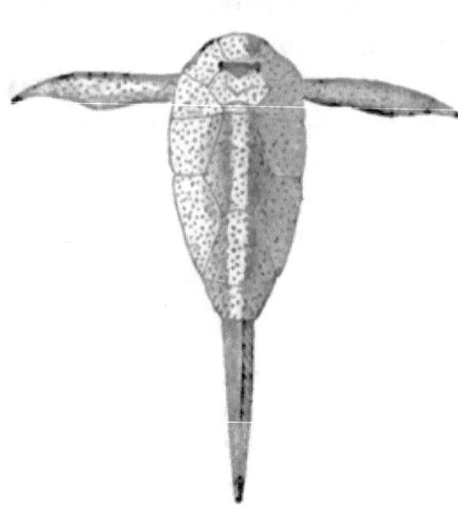

Fig. 39.—Pterichthys.

In their general shape the bodies of fishes exhibit a great compression from side to side, a rounding of the sides, and a pointing of either extremity, adapting the animals for easy progression through the water. Some fishes, such as the soles (fig. 41), flounders, plaice, &c., are named 'flat fishes' from the great flattening exhibited by their bodies; although, at the same time, it is important to observe that these fishes are simply more compressed from side to side than their neighbours. Most persons, on looking at a sole or flounder, are apt to think that one of the flat surfaces must represent the back, and the other the under surface of the body. This idea is strengthened by the fact that the so-called back-surface is dark, and the apparent under surface light in colour, and because both eyes exist on the dark-coloured surface. That, however, the flat surfaces are really the *sides* of the fish may be seen by noting that on each surface a breast-fin is developed; these fins being placed invariably one on each side of the body.

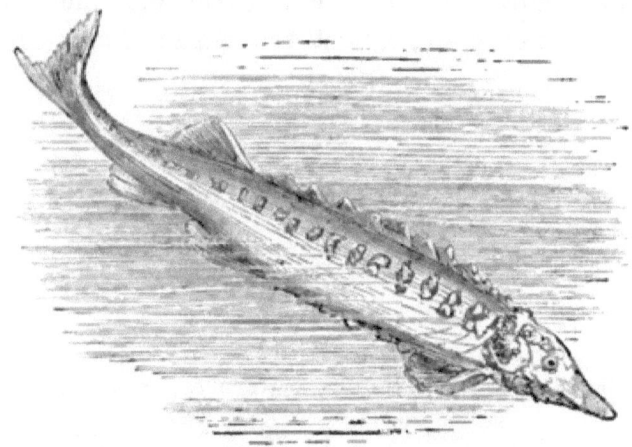

Fig. 40.—Sturgeon (*Accipenser sturio*).

And whilst the eyes in early life are disposed one on each side of the head, in the position in which eyes are naturally situated, they are gradually brought round to one side by the bones of the head becoming curiously twisted in

Fig. 41.—Sole (*Solea vulgaris*).

the course of development. Thus these fishes lie and swim on one side—that which is light-coloured—and present a most singular combination of curious and abnormal features.

The fins of fishes constitute interesting features in their structure. Almost all fishes have two sets of fins—those which exist in pairs, and those which are unpaired, and which are developed in the middle line of the body. To the former class belong the two pectoral or 'breast-fins' (fig. 42, *b*), and the two ventral or 'belly-fins,' *c*. The 'breast-fins' correspond to the fore-legs of other animals or to the arms of man; whilst the ventral fins correspond

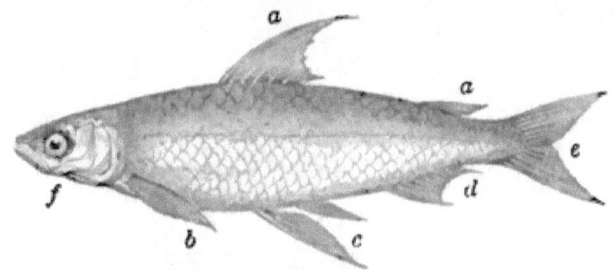

Fig. 42.—Fish shewing the various fins:
a, a, dorsal fins; *b*, pectoral fin of one side; *c*, ventral fins; *d*, anal fin; *e*, caudal or tail-fin; *f*, operculum or gill-cover.

to the hind-legs or to man's lower limbs; and these latter fins may be placed, as hind-limbs should be, to the rear of the body (as in sturgeons); or they may be found (as in the cod) placed beneath the breast-fins on the throat.

It may be asked, How do we know that these two pairs of fins represent the limbs of other animals? We reply, because when we investigate their structure we find them to be supported by a bony skeleton, the various portions of which correspond to those existing in the skeleton of the limbs of man or other vertebrates. And it is only through this important principle of tracing out what are known as the homologies or resemblances between parts, and by looking at and comparing their structure, that we are enabled to find out the real nature of many organs in animals; similar organs frequently existing under very different and varied guises.

The other fins of fishes do not exist in pairs, but are placed in the middle line of the body. Hence they are named the median or unpaired fins. Thus we find the back or dorsal fins, *a*, to represent the unpaired fins, as also do the tail, *e*, and anal, *d*, fins; the latter being placed on the lower surface of the body. These unpaired fins, if they correspond to any other structures in the fishes, are simply to be regarded as special developments of the skin, and therefore bear no true relationship to the limbs of other animals. We may find one or more dorsal and one or more anal fins; but the tail-fin, by the action of which, as every one knows, the fish chiefly swims, is always single,

Fig. 43.—Electrical Eel (*Gymnotus electricus*).

but may be divided into halves. Most of our common fishes (fig. 42) have the halves of the tail-fin of equal size; others, such as the sharks, sturgeons, &c., having the upper half greatly exceeding the lower half of the tail-fin in size. In one species of shark, named the Thresher or Fox-shark, the upper half of the tail-fin appears enormously developed as compared with the lower half; and the names of this species have been derived from the use the fish makes of its tail in lashing the water, and from the long-tailed appearance suggesting a resemblance to the familiar Reynard of the land. In fishes the tail-fin is always placed vertically, or in the same line as the body, and moves from side to side; whilst in the whales—which are not fishes, but Mammalia or quadrupeds possessing fish-like bodies—the tail-fin is placed across the body. Some fishes may want arms or legs—that is, the pectoral or ventral fins; the eels (fig. 43), for example, possessing no ventral fins. The flying-fishes (fig. 44), on the contrary, possess a very large development

Fig. 44.—Flying-fish (*Exocœtus volitans*).

of the pectoral or breast fins, and support themselves temporarily in the air by their aid.

Fishes are usually very well provided in the matter of teeth. What would be thought of a quadruped which had teeth not only in its jaws, but had its tongue, its palate, the sides and floor of its mouth, and other parts, also bearing rows of these structures? Yet such is the case with many fishes. Then, also, where the teeth of one set in fishes are lost, or destroyed through the natural wear and tear to which they are subjected, new teeth are developed to supply the place of the lost members. Any one may gain a good idea of the formidable array of teeth in fishes, and of the manner in which one set succeeds another, by inspecting the jaws of a shark in a museum. In fishes the teeth are not implanted in sockets, but are fastened by ligaments to the surface of the bones which bear them. Sometimes one tooth only, is developed in fishes. This is the case in the curious, eel-like hag-fishes (fig. 45) already mentioned; these fishes possessing but a single large tooth, borne on the palate; and by means of this formidable weapon which possesses saw-like edges, they bore their way into the bodies of other fishes, and there take up their abode as unwelcome guests. A cod or large haddock may

Fig. 45.—Hag.

sometimes be found with five or six hags contained in its interior. The parrot-fishes, or *Scari*, of tropical seas, are so named from their possessing jaws shaped like the beaks of those familiar birds, and these jaws are rendered all the more extraordinary from their being covered or incrusted by numerous small teeth, which are as closely packed on the jaw as paving-stones are in a street, and which serve these fishes as useful instruments when they feed upon the living parts of the hard and limy coral-animals. In the jaws and floor of the mouth of the Port Jackson shark, or in the Eagle rays, or skates, the teeth may be seen to be flat and broad. Such teeth form a regular pavement arranged like a mosaic pattern, and are admirably adapted for crushing whatever substances enter the mouth.

Fishes are well provided in the way of digestive apparatus. A throat or gullet, stomach, intestines, liver, and other glands, serve for the digestion of the food, and a heart and blood-vessels exist for the circulation of the blood thus manufactured from the food. The blood is purified in the gills. Each gill—consisting, in common fishes, of a supporting 'arch' bearing a great number of delicate filaments arranged like the teeth of a comb—may be viewed as simply a network of blood-vessels. The blood, pumped into this network by the heart, is purified by the action of the oxygen gas contained in the pure water which the fish is constantly taking into its gill-chamber by its mouth; whilst the pure blood is re-circulated through the body, and the water used in breathing is got rid of by being ejected behind the 'gill-cover' (fig. 42, f) at the neck, so as to allow a fresh inflow to be drawn in by the mouth. The gills of some fishes may be very differently constructed from those of the common members of the class. Thus the lampreys breathe by pouch-like gills which open each by a separate aperture. Seven gill-apertures may be seen on each side of the neck of the common Lamprey (fig. 46); and the sharks, skates, and their

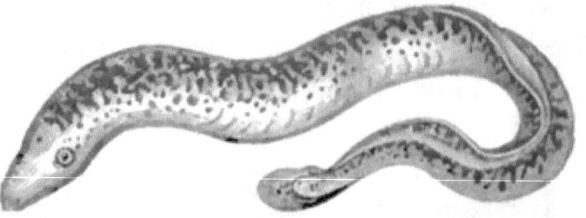

Fig. 46.—Common Lamprey (*Petromyzon marinus*).

neighbours also breathe by sac-like gills. Certain curious facts regarding the breathing of fishes will be afterwards alluded to. Fishes illustrate plainly what is meant by aquatic or water-breathing. They possess gills or organs, adapted for separating the atmospheric air which is entangled or contained in the water; land-animals breathing the same air directly from the atmosphere.

That fishes are wary and active, and possess senses of acute nature, are facts well known to all. The lowest fish, the little clear-bodied Lancelet (fig. 47), possesses no brain whatever, and no organ of hearing is developed, whilst the eyes are at the best of very simple and rudimentary structure. In other fishes, again, the brain and nervous system not only acquire a typical development, but the senses also advance in perfection. The sense of sight is of perfect kind, the eyes of fishes being adapted for seeing in the dense medium in which they live; whilst the sense of smell is also developed, although, curiously enough, the nostrils, in all except two

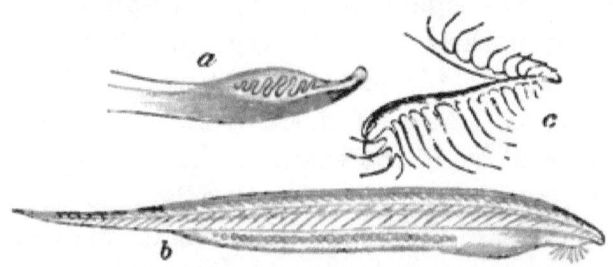

Fig. 47.—Lancelet (*Amphioxus lanceolatus*):
a, mouth, seen from below; *b*, general view of body; *c*, hyoid bone, with mouth-filaments.

kinds of fishes—the hag-fishes and the curious *Lepidosiren* or mud-fish—are pocket-like in nature, and do not open backwards, as in higher animals, into the mouth. The sense of taste is not exercised in a high degree by fishes, and it is interesting to observe that the sense of touch appears to reside especially in the sides of the body, on which surfaces a well-marked line—the 'lateral line'—may be observed in most fishes. This lateral line is connected with a series of canals or sacs abundantly supplied with nerves. The function of these organs is believed to be that of exercising the sense of touch; and from the manner in which many fishes swim against objects, and bring the sides of their bodies in gentle contact with foreign objects, there would seem to exist strong reasons for supporting the above idea. That fishes 'hear' is a well-known fact. No outer ear is developed, but an internal ear—the essential part of the organ of hearing—is found in all fishes except the little Lancelet.

Whilst the intelligence or instinct of fishes is not, generally speaking, of a high order, there are not wanting instances to prove that these animals may exhibit traits of character sometimes wanting in higher groups of animals. Any one who has kept gold-fishes must have noted that in time they become more and more familiar with the hand that feeds them, and the experience of aquarium-keepers goes to prove that some fishes may even shew signs of recognising friends. One of the most remarkable incidents illustrating the latter fact is related of a ling. As a Greenock diver, some years ago, was engaged in working at the wreck of the *Duke of Edinburgh* at Ailsa Craig, he became aware that he was an object of interest to a fish of the ling species, which had observed him at work. On the first day it did not come within six feet of where he stood, and took alarm when any of the men in the lighter above looked over the side. The wreck lay at a depth of ten fathoms, and the men in attendance could clearly see the diver at work. On the second day the fish, which was about three feet in length, came quite close, swam round about the diver, and would poise itself for ten or fifteen minutes at a time, and watch his movements. As time wore on, it became more familiar, and amused itself by swimming round his head. It would then shoot upward through the line of air-bubbles that emerged from the escape-valve till it reached the surface, when it gave its caudal fin a flourish in the air, and descended again through the air-bubbles to the bottom, continuing these freaks sometimes for an hour without intermission. The third day was stormy, rendering it impossible for the diver to descend, but in the forenoon heavy rain came on, and the sea fell. The diver then descended to his work, taking some bread with him to the depths below. The fish came and ate the bread from his hand, and latterly became so tame that it allowed him to touch it. This friendly intercourse continued till the sixth day, when one of the men in attendance, who had at night thrown a fishing-line over the side, pulled in the diver's pet, and not seeing any difference between it and any other ling, had it cleaned and prepared for food. The sad termination to the narrative fortunately does not lessen the interest in the tale, which thus appears strongly to argue in favour of the possession by some fishes of a high degree of intelligence.

Like the human race, the class of fishes evinces many illustrations of individuals and groups which differ more or less widely from their more commonplace neighbours. To some of the more curious of these 'odd fishes' we may next direct attention.

A very singular little group of fishes, for example, is that known to the naturalist by the name *Lophobranchii*; this term meaning literally 'tuft-gilled.' Included in this division are two curious families, of one of which the Sea-horses (fig. 48) or *Hippocampi* are the representatives; whilst to the other family belong their allies, the Pipe-fishes. No more interesting forms than these two groups can well be selected from the great class of which they are little-known members. And the interest with which they are regarded by zoologists extends beyond the

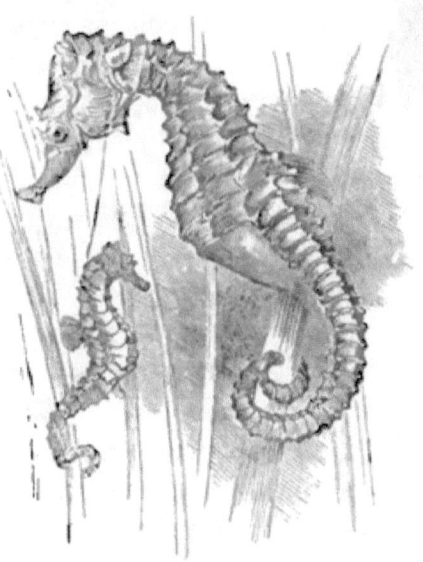

Fig. 48.—Sea-horses.

mere investigation of their outside form or appearance; since they present, in many points of their economy and habits, very marked deviations from what one may call the ordinary course of fish-life.

To visitors to the great aquaria which are now springing up in every part of the country, the sea-horses will be familiar. Their hardy nature together with their curious appearance have marked them out as aquarium favourites; and they may fairly, in respect of their zoological fame, divide the honours with any of their companion-tenants. Imagine a little body from four to six inches in length, topped by a head which in outline exactly resembles that of a horse, and which tapers off below, or rather behind, into a lithe, flexible, and pointed tail, and we may form a rough idea of the general appearance of one of the sea-horses. This little body we shall find to be covered with ganoid plates or scales of hard horny or bony material, exhibiting ridges and angles all over its surface. Two large brilliant eyes, each of which may be moved independently of the other, add to the curious appearance of the head; whilst to the body itself may be attached long streamers of sea-weed, serving to conceal the little beings as they nestle amid their marine bowers, each looking like some veritable creation of heraldic or mythological kind.

The flexible tail which terminates the body has the important office of mooring or attaching the fishes to any fixed object. As we see them in the aquarium, they are generally poised, as it were, on the tail; the latter being coiled around a bit of sea-weed, whilst the erect body and head look warily through the waters of their miniature sea. When they detach themselves, they swim about in the erect position by means of the two pectoral or breast fins, which being placed close to the sides of the neck, project like veritable ears, and assist in rendering the equine appearance of the head of still more realistic nature. These fins move with a quick twittering motion, and propel their possessor swiftly through the water; whilst the back-fin, placed towards the hinder extremity of the body, also assists them in swimming.

Some curious points in the internal structure of the sea-horses warrant a brief notice. As already stated, the gills of an ordinary fish are shaped each like a comb; the teeth of the comb being represented by the delicate processes, each consisting in reality of a network of blood-vessels, in which the blood is exposed to the oxygen of the water, and is thus purified. In the sea-horses, however, the gills do not present this comb-like appearance, but exist in the form of separated tufts or bunches of delicate filaments, which spring from the gill-supports or arches. From this peculiarity, the name 'tuft-gilled,' already alluded to, is derived, and the pipe-fishes agree in the structure of the gills with the sea-horses. Then, also, as most readers are aware, the gills of ordinary fishes are covered by a horny plate, appropriately named the gill-cover (fig. 42, f), and it is by sharply compressing the gills with this cover, that the water used in breathing is ejected from the gills, so as to make room for a fresh supply. In the sea-horses, however, the gill-cover is not open or free at its under and hinder edges, but is firmly attached all round to the neighbouring tissues, and so rendered immovable. At one point in its circumference, however, a small aperture is left, through which the breathing-water escapes from the gills.

The sea-horses are found abundantly in the English Channel, around the coasts of France and Spain, in the Mediterranean Sea, and in the tropical oceans. Several distinct species are known to zoologists, but they closely resemble one another in the essential features just noted. They are lively and intelligent little creatures, and become familiar in time with their possessors. Fixed by their tails, they may be seen actively to dart the head at any passing object adapted for food; whilst, when they wish to free their bodies from the attached position, they appear to manœuvre with the chin and head in order to effect their purpose. Their food appears to consist of small crustaceans, worms, &c.; and they are known to be especially fond of such delicate titbits as are afforded by the eggs of other fishes.

Perhaps the most curious part of the history of the sea-horses relates to their care of the young. Fishes generally take little or no care of their offspring, and it is therefore the more surprising to encounter in the little beings before us a singular example of parental fidelity and attachment. Nor, as might be expected, is it the mother-fish who is charged with the task of attending the young. Contrary to the general rule, the male fish assumes the part of nurse, and well and faithfully does he appear to discharge his duties. At the root of the tail in the male sea-horses, a curious little pouch is seen. In this pouch the eggs laid by the females—which want the pouch—are deposited, and are therein duly hatched. Nor does the parental duty end here; for after the young are hatched and swim about by themselves, they seek refuge in the pouch during the early or infantile period of their life whenever danger threatens them. This procedure forcibly reminds one of the analogous habits of the kangaroos and their young; but the occurrence is the more remarkable in the lower and presumably less intelligent fish.

Some experiments made on the sea-horses seem to demonstrate the existence of a more than ordinary degree of attachment to the young. Thus when a parent-fish was taken out of the water, the young escaped from the pouch; but on the parent being held over the side of the boat, the young at once swam towards him, and re-entered the pouch without hesitation. Some authorities have not hesitated to express an opinion that the young are nourished within the pouch by some fluid or secretion of its lining membrane. But further observation is certainly necessary before this latter opinion can be relied upon.

The Pipe-fishes are very near neighbours of the sea-horses, and derive their name from the thin elongated shape of their bodies, together with the fact that the jaws are prolonged to form a long pipe-like snout, at the extremity of which the mouth opens. These fishes are very lively in all their movements, and dart through the water so quickly, that in many cases the eye is unable to follow them. Like the sea-horses, the male pipe-fishes protect and tend their progeny, and exhibit an equal attachment to their young.

These latter features are also well exemplified by the familiar Sticklebacks of our ponds and streams. The latter fishes actually build nests for the reception and care of their eggs, the nests being made chiefly or solely by the males; whilst on the latter, during the process of hatching and in the upbringing of the young, devolves the chief care of protecting and looking after the welfare of the progeny. These instances of the care and duties which devolve on the males, instead of on the mother-parents, appear to reverse the more natural order which almost universally obtains in the case of both lower and higher animals.

Of the oddities which fish-life presents, probably none are more remarkable than the Archer or Shooter Fishes (*Toxotes*), which inhabit the seas of Japan and of the Eastern Archipelago. When kept in confinement, these fishes may be seen to shoot drops of water from their elongated jaws at flies and other insects which attract their attention. They have been observed to strike their prey with unerring aim at distances of three or four feet. Another notable species of shooting-fishes is the *Chætodon*.

This latter form possesses a prominent beak or muzzle, consisting of the elongated jaws; and from this beak, as from the barrel of a rifle, the fish shoots its watery missiles at the insects which alight on the vegetation fringing its native waters.

The old saying which compares great helplessness to the state of 'a fish out of water,' does not always find a corroborative re-echo in natural history science. As every one knows, different fishes exhibit very varying degrees of tenacity of life when removed from their native element. Thus a herring dies almost immediately on being taken out of water; whilst, on the other hand, the slippery eels will bear removal from their habitat for twenty-four hours or longer; and we have known of blennies—such as the Shanny (*Blennius pholis*)—surviving a long journey by post, of some forty-eight hours' duration, when packed amid some damp sea-weed in a box.

But certain fishes are known not merely to live when taken out of water, but actually of themselves, and as part of their life and habits, to voluntarily leave the water, and disport themselves on land. Of such abnormal fishes, the most famous is the Climbing Perch or *Anabas scandens* (fig. 49) of India, which inhabits the Ganges, and is also

Fig. 49.—Climbing Perch.

found in other Asiatic ponds and rivers. These fishes may be seen to leave the water, and to make their way overland, supporting themselves in their jerking gait by means of their strong spiny fins. They appear to migrate from one pool to another in search of 'pastures new,' especially in the dry season, and when the water of their habitats becomes shallow.

The Hindu name applied to these fishes means 'climbers of trees;' and although statements have been made both by travellers and natives, that the climbing perch has been found scaling the stems of trees, these accounts, we fear, must be regarded as of equal value with the native belief that the fishes fall in showers on the land from the skies. Of the power of the fishes to live for five or six days out of water, however, no doubt can be entertained; and their ability to support life under these unwonted conditions is explained by the fact that certain bones of the head are curiously contorted so as to form a labyrinth, amid the delicate recesses of which a supply of water is retained, for the purpose of keeping the gills moist.

Another group of fishes, also inhabiting India, and possessing powers of existing 'out of water,' is the *Ophiocephalidæ* ('snake-headed'); a family allied to the Mullet-group. It would appear, from some recent observations on these fishes, that they are enabled not only to live, like the climbing perch, out of water, but that they die if kept below the surface of the water even for a comparatively short time. Thus when an Ophiocephalus and a carp were placed together in a vessel of water, a net being placed about two inches from the surface, the carp swam, as might be expected, freely and continuously below the surface; whilst the Ophiocephalus made vigorous efforts to attain the surface, for the purpose of inhaling air directly from the atmosphere. When not allowed to reach the surface, the Ophiocephali died, suffocated, in periods varying from twenty minutes to two hours. A carp secured by a bandage placed round the gill-slits, so as to prevent any escape of water by these apertures, died, as might have been expected; whilst an Ophiocephalus treated in the same manner exhibited no symptoms of uneasiness. The explanation of the power possessed by the latter fish of being able to live out of water, resides in the fact that these fishes possess two cavities in the throat, in which blood is purified by the inhalation of atmospheric air. Thus, the Ophiocephalus not only can exist out of water, but escape from that medium must, in fact, be viewed as an absolute necessity for the normal life of the animal. The climbing perch appears also to exhibit this latter peculiarity of requiring to escape periodically from the water, for this fish, like the Ophiocephalus, may be actually drowned, if kept from obtaining a supply of atmospheric air.

The curious *Lepidosirens* or Mud-fishes, which occur in the Gambia of Africa and the Amazon of South America,

Fig. 50.—Lepidosiren (*Lepidosiren annectans*).

exhibit a greater peculiarity of structure, which still more completely fits them for living out of water. In the great majority of fishes, a curious sac or bag known as the *swimming* or *air bladder* is found. The use of this structure in ordinary fishes is to alter the specific gravity of the animals; and, by the compression or expansion of the air or gases it contains, to enable them to sink or rise in the water at will; but it would also appear that indirectly it may aid in the breathing of all fishes which possess the organ. In the mud-fishes, however, the air-bladder becomes divided externally into two sacs, whilst internally each sac exhibits a cellular structure resembling that seen in the lungs of higher animals, with which structures, in fact, the swimming-bladder of fishes actually corresponds. Then also this elaborate air-bladder of the mud-fish communicates with the mouth and throat by a tube, which corresponds to a windpipe. The nostrils of the mud-fishes further open backwards into the mouth; whilst, as already mentioned, in all other fishes, save one genus, the nostrils are simple, closed, pocket-like cavities. And it may lastly be noted that the *Lepidosirens* are in addition provided with true gills, like their ordinary and more commonplace neighbours.

These remarks serve to explain the 'reason why' these fishes can exist for months out of water. Thus, on the approach of the hot season, the mud-fishes leave their watery homes, and wriggle into the soft mud of their native rivers. Here they burrow out a kind of nest, coiling head and tail together; and as the mud dries and hardens, the fishes remain in this temporary tomb; breathing throughout the warm season like true land-dwellers, by means of the lung-like air-bladder. When the wet season once more returns, the fishes are aroused from their semi-torpid state by the early rains moistening the surrounding clay; and when the pools and rivers once more attain their wonted depth, the *Lepidosirens* emerge from their nests, seek the water, breathe by means of their gills, and otherwise lead a true aquatic existence. Another fish, the *Ceratodus* or 'Barramunda' of Australian rivers, possesses a similarly modified air-bladder, and is thus enabled to breathe independently of its gills.

With such a combination of the characters of land and water animals, it is little to be wondered at that the true position of the mud-fishes and their neighbours in the zoological scale should have formed a subject for much discussion. They appear, however, to be true fishes, and not amphibians (or frog-like animals); and they therefore may legally occupy a prominent position among the oddities of their class.

Other curious beings included among the fishes are the so-called Globe-fishes (*Diodon*, &c.), which derive their name from their power of distending their bodies with air at will; and their bodies being usually provided with spines,

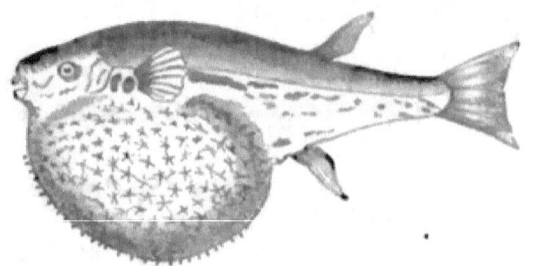

Fig. 51.—Globe-fish (*Tetraodon lævigatus*).

they may be judged to present a rather formidable front to any ordinary adversary in their expanded condition. Then also we have the curious Trigger-fishes (*Balistes*), so named from the prominent pointed spine in front of the first back-fin; this spine firmly holding its erect position until the second spine or fin-ray be depressed, when the first spine is released by mechanism resembling that of the trigger of a gun. The obvious use of such an apparatus is clearly of a defensive kind; and it is remarkable to find that man has imitated and reproduced, in one of his common mechanical contrivances, a structure existing in all its natural perfection in the fish.

Oddities in the way of curious fishes can receive no better illustration than that afforded by the very curious 'Telescope-fishes' of China, figured in the accompanying illustration. These beautiful little fishes are kept alive in many of our large aquaria. At first sight, the telescope-fishes might be mistaken for the familiar gold-fishes, but a cursory inspection of their appearance at once shews the peculiarities of structure which have earned for these creatures their distinctive name. The eyes are seen to be singularly prominent, and protrude from the head to a marked extent, whilst they also present certain alterations in intimate structure. The fins, moreover, are double, this conformation being well exemplified in the large and prominent tail-fin. The exact nature of these fishes has been discussed by the French Academy of Sciences, in the records of which it is stated, that the Chinese have cultivated these fishes

Fig. 52.—Telescope-fishes.

from an ordinary species of the carp race, and that the peculiar conformation of the eyes results from a diseased state, which, by being transmitted from one generation to another, has become at last a stable and definite character of the animals. This very probable explanation of the origin of these peculiar eyes, is supported by the fact that certain carps inhabiting the canal Saint Martin at Paris were found to possess prominent eyes; and a like appearance has been observed in carps living in rivers into which the water of drains had been allowed to flow. The curious fact is thus brought under notice, that a diseased condition may not only entirely alter the appearance of an animal, but may be faithfully reproduced in its descendants.

As a final feature of interest in the history of fishes, we may allude to the large size attained by certain species. The huge sharks and their allies may attain an immense size; but some fishes, usually of moderate or small size, may occasionally grow to an unusual extent. Thus the mackerel, the ordinary length of which is not above thirteen or fourteen inches, has been known to attain a length of nearly two feet and a weight of two pounds six ounces; and the halibut has been known to measure seven feet in length, three and a half feet in breadth, and to weigh between two and three hundred pounds. Cod have been met with measuring over nine feet in length; one specimen captured on the west coast of Scotland in 1877, having measured nine feet two and a half inches in length, and three feet two and a half inches in circumference. And one of the Rays—the Horned Ray of the Mediterranean—is known frequently to attain a length of twenty feet, a breadth of twenty-eight feet, and a weight exceeding one ton. These giants of their race, no less than their stranger neighbours, already noticed, may be regarded as exemplifying in a singular manner the curious and strange in fish-existence.

Fig. 53.—Nest of Termite Ant.

SOME CURIOSITIES OF INSECT-LIFE.

ALL who watch with any degree of attention the growth of plant-life in a garden or greenhouse must be familiar with those curious little green insects, the Aphides or Plant-lice. Existing in thousands on our flowers and shrubs, and feeding on the juices of the plants, they constitute veritable pests; and some species, infesting the bean, hop, and other cultivated plants, cause much anxiety to the agriculturist from their destructive effects on his crops. Both sexes of aphides are generally found in a wingless state; although, as will presently be explained, the individuals of the same species may possess wings at one period, and be wingless during the rest of the year.

A fact of primary interest in the habits of these insects consists in the attentions paid to them by the familiar ants —the famous Huber being the first to make this observation. Thus the ants may be observed to follow the plant-lice, and to stroke the abdomens of the latter with their antennæ or 'feelers;' this act causing the aphides to exude a sweet viscid secretion from two tubular pores placed towards the hinder extremity of their bodies. This secretion is greedily absorbed by the ants. Mr Darwin mentions an observation of his own which seems to strengthen the idea that the relations between the ants and their providers are

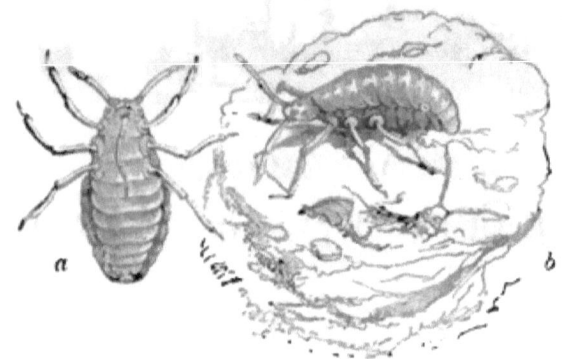

Fig. 54.—Apple Aphis (*Eriosoma mali*):
a, wingless insect (magnified); *b*, wingless insect in excrescence of the tree (magnified).

of a very intimate and reciprocal kind. Having removed all the attendant ants from a group of about a dozen plant-lice which resided on a dock-plant, Mr Darwin prevented the ants from regaining their vantage-ground for several hours. Feeling certain that the aphides would by that time have secreted a goodly store of the sweet secretion of which the ants are so fond, Mr Darwin watched them intently for some time, but did not observe a single aphis emit the secretion. He then tried to imitate the movements of the ants' antennæ by stroking the abdomens of the plant-lice with a hair; not a single aphis, however, responding to the imitative demand. A single ant admitted to the guarded aphides, was observed to hurry from one to the other, as if aware of the plentiful store of sweets awaiting its attention; and when this single marauder, if we may so term it, began to stroke the various plant-lice with its antennæ, the latter rapidly excreted the coveted fluid, the secretion being greedily absorbed by the ant. Very young plant-lice similarly respond to the call of their ant-neighbours; and this latter fact would tend to shew the purely instinctive and hereditary nature of the curious impulse on the part of the aphides; whilst the action of the ants in the matter must be no less clearly of instinctive kind.

But exceeding in interest even the curious habits just noted, we find the development of the aphides to present us with some phases of puzzling and inexplicable aspect. At the close of autumn, male and female aphides are found living on plants. The eggs produced by these insects, after lying dormant throughout the winter season, burst into active life in the succeeding spring, and give birth, not to males and females, as might be expected, but to wingless, six-legged plant-lice, which, if their sex be determinable at all, must be regarded as belonging to the female sex.

Now appear some curious phenomena; for if these wingless females be watched, they may be seen to produce, alive, and not from eggs, brood after brood of young plant-lice, exactly resembling themselves, in that they wholly consist of female insects, and resembling their parents, in that they are destitute of wings. Throughout the spring, summer, and autumn, each successive generation of these wingless females thus produces progeny which repeat the features of their spinster-like parents; not a single individual of the 'sterner' sex being found within the limits of this Amazonian population. This uninterrupted succession of female generations may be repeated and traced in a single season, through nine, ten, or even eleven generations; whilst the number of the progeny of a single aphis-mother may amount, as estimated by Reaumur, to 5,904,900,000 at the fifth generation alone. At length, when the close of autumn once more comes round, and ten or eleven generations have been born, this uninterrupted succession of female progeny ceases, and in the last brood winged males and females appear—as at the similar period of the preceding year, when our survey of their life was supposed to begin. Then, as before, eggs are produced by this last generation; and from these eggs, in the succeeding spring, will be developed the wingless females, whose descendants will repeat the strange history of the preceding year.

If we appeal to the zoological world for an explanation of these curious facts, we shall find that several conflicting theories and opposing views prevail. As all must admit, the circumstances above detailed, and as verified by repeated observation, leave no doubt on the mind that the ordinary laws of development are not only set aside, but are incompetent of themselves to aid us in obtaining a solution of the matter. Thus, it has been supposed that the reproductive influence of the original and ordinary development, through eggs, of the first brood of the male and female aphides, extends throughout the succeeding generations. This, however, is merely a theoretical possibility, and does not aid us in the explanation of the anomalous fact, that one sex alone is enabled to produce living progeny; whilst under ordinary circumstances, and throughout the whole range of the animal world, the co-operation of both sexes is necessary to develop eggs capable of evolving progeny. The case of the plant-lice would be paralleled by that of plants, if the seeds of plants furnished by the pistils could be duly fertilised and be rendered capable of developing a new plant, without the influence of the necessary pollen-substance furnished by the stamens. And this, as far as we know, rarely, if ever, occurs.

Naturalists know these phenomena under the name of *parthenogenesis*; and probably the best explanation of the development of the plant-lice, together with allied cases in other insects, is that the eggs resemble 'buds' in their essential nature; and whilst eggs ordinarily require for their development the presence of both sexes, the generations of female aphides may be regarded as produced from their single parents, by a process of internal budding. The stock or structure in the females from which these egg-buds are produced may be named 'germ stocks'—the *Keimstöcke* of the German naturalists. And in this view, we might not inaptly compare the life-cycle of the aphides with that of a plant. The plant springs from the fertilised seed—as do the original aphides from true eggs. The plant further by budding produces through the greater part of the year its leaves and other organs—as the spinster-aphides produce their young by an analogous process. Then, in due season, the stamens and pistil, or reproductive organs of the plant, are formed, and true seeds capable of giving origin to a new plant are produced—just, indeed, as the aphides in their turn develop both sexes, and as, from the eggs thus produced, new beings with special powers and tendencies are introduced into the wondrous cycle of their life. The consideration of such interesting phenomena as the preceding, should forcibly impress us, above all other considerations, with the marvellous plasticity of living forms, and with the endless variety of contrivance and action which, for the accomplishment of its own duly arranged ends, life is continually exhibiting before us.

Not less interesting than the preceding facts are the details regarding the life and habits of ants, with which the observations of many naturalists have made us familiar. If some of the ancient ideas concerning the provident and industrious habits of these insects have been ascertained to be erroneous, modern science has on the other hand revealed phases of ant-life far exceeding the older ideas in wonder and interest. Thus certain kinds of ants are well known to be slave-makers; that is, they seize upon the young of other kinds, and convey them captives to their nests, there to train them up as veritable servitors. Thus one species (*Formica sanguinea*) makes raids upon the 'negro' ants, and carries off the young 'negroes' alone; the latter, as they grow, coming in time to perform all the work of their masters' home. Being trained literally from their youth upward, the negro-ants, as might be expected, make excellent servitors. Another slave-making species of ant, the Amazon-ant (*Polyergus rufescens*) appears to be forced to depend on servitors for help in the management of the nest; since the mouths of the masters in this case are so constructed that, as a famous naturalist observes, 'it is physically impossible for the rufescent ants (or masters), on account of the form of their jaws, and the accessory parts of their mouth, to prepare habitations for their family, to procure food, or to feed them.' Huber found, indeed, that this species, when left without slaves, perished from sheer inability to help themselves. This observer placed thirty of these slave-making or Amazon ants with some of their own young (pupæ and larvæ) in a box, along with some honey. 'At first,' Huber remarks, 'they appeared to pay some little attention to the larvæ; they carried them here and there, but presently replaced them. More than

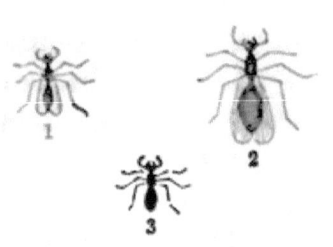

Fig. 55.—Ants:
1, Female; 2, Male; 3, Neuter (Worker).

one half of the Amazons died of hunger in less than two days. They had not even traced out a dwelling, and the few ants still in existence were languid and without strength. I commiserated their condition, and gave them one of their black companions. This individual, unassisted, established order, formed a chamber in the earth, gathered together the larvæ, extricated several young ants that were ready to quit the condition of pupæ, and preserved the life of the remaining Amazons.' Thus the Amazon-ants actually perish when their slaves are not at hand to feed them. No more typical instance of the interdependence of animal forms could be found; and the facts thus presented become perhaps the more wonderful, if we reflect that the Amazons, whatever may have been their original capacities, have to all appearance lost even the instinct of feeding. They are thus literally tended 'hand and foot' by their slaves, and present sad examples of a want of training in the useful arts and accomplishments of domestic life.

Foreign species of ants are known to exhibit engineering habits, and to display great skill, in overcoming natural obstacles which may lie in the line of march. The Driver-ants (*Anomma*) of Western Africa march in hordes, and devastate the country through which they pass. They cross streams of considerable breadth by forming living bridges of interlocked bodies, over which the moving army passes—this procedure, curiously enough, being imitated by certain species of monkeys. And other equally interesting facts have been noted by observers regarding tropical species of ants; as for example, where obstacles purposely thrown in the line of march have either been skilfully removed, or have been circumvented by plans and manœuvres exactly resembling those practised in analogous circumstances by mankind.

A Leaf-cutting Ant (*Œcodoma*) carries leaves for long

distances, and detaches distinct parties of its members for special work, just as the divisions of an army or regiment may be allotted to distinct duties. Thus foragers are told off to climb the trees and cut off the leaves; whilst carriers take up the fallen leaves and transport them to the nest. Here we find a perfect example of the economical subdivision of labour.

A series of most important and interesting observations on the intelligence of ants has recently been detailed by Sir John Lubbock; these investigations shewing that the instincts of ants, however wonderful they may appear, are after all devoid of the higher or reasoning power seen in man himself, and that in certain cases a decided inaptitude to overcome difficulties by a ready and simple method is exhibited by these insects. For example, some food was placed on a slip of glass in a cup, the food being surrounded by water, which, however, was bridged over by means of a strip of paper two-thirds of an inch long, and one-third of an inch in width. An ant (*Formica nigra*), on being shewn the food, set to work to convey it to the nest, several others being soon brought to assist in the work. When about twenty-five ants were thus employed, the paper-bridge was slightly moved so as to leave a little gap, separating the food from the bridge. When the ants came to this chasm they made every effort to get over, but after repeated efforts, gave up the attempt and returned home fairly beaten. Sir John Lubbock remarks that it never occurred to the ants to move or to push the bridge, although the bridge itself was so light, and the gap only one-third of an inch wide. A piece of straw being substituted for the paper-bridge, on the idea that the former substance was one with which the ants might be more familiar, the insects exhibited the same incapacity to overcome this apparently simple difficulty. Food being placed just above their nests, the insects invariably carried the food to their homes, and did not on one occasion save time and trouble by dropping the food on to the nest. When food was placed at a distance of one-third of an inch above the nest, the ants tried to reach down, but neither did they jump down themselves, nor did they throw down the food, although the interval was of so small extent that when an ant passed below the food, the ant above could descend by stepping on the back of its neighbour. This latter mode of descent only happened accidentally. A heap of fine mould was placed close to the glass on which the food was set; and although the ants, by moving a particle of earth for a quarter of an inch, could have made a bridge across which the food could readily have been conveyed, not a single insect appeared to 'think' of such a contrivance. Yet in contrast to this seeming stupidity, must be placed the incident in which food was placed in a shallow box fitted with a glass top, and provided with a hole on one side. The ants being introduced to this food, carried it off to the nest through the hole in the box. Some fine mould was then placed in front of the hole, so as to cover the aperture with earth to the extent of half an inch. The ants which were in the inside of the box were then removed, and along with their neighbours were allowed to seek their food. After searching for some other entrance to the box, and finding none, they excavated with care and precision into the mould which obstructed the former entrance, and thus obtained admittance to the store. It is reasonable to suppose, however, that the exercise of intelligence in the latter instance may be explained on the ground that the obstruction was of a kind and nature which the ants are accustomed to encounter in their ordinary and natural state of existence. Collections of earth and like *débris* are obstacles which may reasonably be supposed to be of frequent occurrence in the ordinary life and operations of these insects, and the work of excavation constitutes a labour in which ants must frequently engage. The fording of miniature streams, as in the previous experiment, on the other hand, is a work not so familiar to these insects; and hence we may argue that custom and habit possess as great an influence over the regulation of their life as over that of the highest of animals.

Various authors have stated that ants will rescue a comrade who is in danger, and thus prove literal friends in need to distressed companions. Sir John Lubbock found by experiments on two species at least, that these observations do not apply, at anyrate, to ants as a whole. An ant buried under earth over which its companions walked to and fro in their march to obtain food, was passed unnoticed and unheeded in two instances. Some very curious facts, however, were ascertained regarding the degree of attention paid by ants to members of their own households and to stranger-ants respectively. When five ants belonging to a particular nest and five others from another nest, but of the same species, were chloroformed and placed near some honey surrounded by water, and on the path which led to the food and which was traversed by their sensible neighbours, the stupefied ants were allowed to lie unheeded by the passers-by for more than an hour; not a single ant appearing to perform the part of the Good Samaritan. Indeed, so far from any charitable feelings being evinced by the healthy towards the sick, we are told that one of the strangers was picked up and dropped into the neighbouring water. Then a member of the family-circle was treated in the same unceremonious fashion, and by-and-by the other sick persons were placed in the fluid. One of the stranger-ants was taken into the 'nest, but for some reason or other the patient was brought out in about half an hour, and thrown into the water. The chloroformed ants being in reality dead—since these insects do not recover from chloroform—Sir John Lubbock caused some ants, including strangers to and natives of a particular nest, to become intoxicated. The sober ants, we are told, removed twenty-five of their friends and thirty of the strangers. Twenty of the friends were carried into the nest and were seen no more; the remaining five were thrown into the water; whilst of the strangers, twenty-four met the latter fate, six being taken into the nest; but of these six, four were brought out and left to themselves, only two being retained. In this latter case, the friends were certainly treated in a more becoming fashion than the strangers, but the code of ant-morality does not altogether appear to be of a very high type—although, indeed, we must not presume to judge the ant from a too strictly human, and possibly short-sighted point of view.

Of the care with which many insects provide for the welfare of the future young, and of the marvellous instincts which sometimes lead a young insect to seek the surroundings which are suitable and necessary for its development, we find a most interesting example in the case of a beetle, Sitaris by name. Sitaris is a near neighbour of the blister-beetles, and lives as a parasite on certain kinds of bees. The eggs of the beetle are laid at the entrance to the bees' nests, and are hatched in autumn. The young beetles remain dormant until the succeeding April, when they become active, and their first act is to settle themselves upon the male bees—which are developed before the females—as they emerge from their nests. This step is but preparatory to another, that of passing from the male bees to the females; and this latter proceeding seems to evince a degree of instinct and cunning of most surprising kind, since the object and aim of the young beetles is that of gaining access to the eggs of the bees. The immature beetles thus pounce upon the eggs as the females deposit them, and each beetle, fastened up by the bee along with honey in the cell containing the egg, becomes thus regularly imprisoned, like a concealed enemy in a neighbour's house. The beetle-larva begins operations by devouring the egg of the bee, a process occupying the first eight days of its imprisonment; after which period it grows largely and floats passively on the surface of the honey, mouth downwards. In this convenient position it consumes the sweet store intended by the mother-bee for the nourishment of her already-devoured progeny. The process of moulting meanwhile proceeds apace in the beetle-larva, and this curious life-history in one sense terminates with the appearance of the perfect beetle in the month of August.

ON SOME CURIOUS ANIMAL COMPANIONSHIPS.

IF it be true, as the old proverb informs us, that 'Poverty makes us acquainted with strange bedfellows,' so no less truly may it be asserted, that natural history science exemplifies for us instances of the strangest associations and companionships amongst both lower and higher animals. Nor are these associations always to be explained on the grounds of parasitism, or from other causes which zoology may plainly enough demonstrate. In cases where one animal acts the part of an unconscious or unwilling 'host' to other animals, which have taken up their abode within or upon it as 'guests,' the cause or principle of the association is quite explicable, on the ground that the parasites seek the bodies of other animals as their natural and rightful territory. And indeed, unless provided for, by gaining access to its own and generally limited territory, the parasite perishes, being literally unable to help itself.

The instances of companionship to which we specially refer, however, are very far removed in their essential features from the question of parasitism. Abundant examples, as we shall presently note, may be found, in which one animal form associates itself with another, often of widely different **nature and status in the scale** of being from itself; this **association being** generally of the **most** invariable kind. The one animal being found, we may safely and surely predict the presence of the other. Such instances of invariable **and close companionship are very rarely to be explained on ordinary grounds, and present to the naturalist puzzles of the gravest and deepest kind. In the vast majority of cases, he fails to see any apparent benefit or aid to be derived by either of the associated beings; and it is exactly this want of object, if we may so term it, in the** companionship **of many animals, which forms one of the most** inexplicable **aspects of such studies.**

It is a remarkable **fact that an absolute disinterestedness marks** many such companionships, **although** it is sometimes **hard to draw the line** which shall separate **pure 'parasites' from mere 'guests' and '**lodgers.**' The** well-known flower-like **sea-anemones, so familiar as denizens of our sea-coasts, and which have been described in a previous article, present several notable examples of curious companionship. It has been noted that small fishes are frequently in the habit of swimming about within the mouths and** stomach-sacs **of large anemones inhabiting tropical seas,** evidently on the **best of terms** with **their hosts. And** this association may be shewn **to be** rather inexplicable, in one sense at least, if we consider that the slightest touch is usually sufficient to cause the tentacles and mouth **of sea-anemones to** close upon foreign objects. **Unfortunate crabs, for example, which** chance **in** their peregrinations **to stumble** against **a** large **sea-anemone, are quickly drawn into the** mouth **by the tentacles and swallowed. Noting this** very **natural feature of anemone-character, it seems curious to** think **of such a** dainty **morsel as a fish being** permitted to **swim at its ease** literally **within the stomach-sac, and** within easy and tempting reach **of its strange neighbour.**

But this very kind of association evinces further curious characteristics; for observers have noted a little fish that not only lives within the Dahlia Wartlet Sea-anemone, but actually permits the anemone to contract itself, and to enclose it in its fleshy tomb without injury. Another sea-anemone—the *Adamsia palliata*—the pretty little 'Cloak-anemone' of our English coasts, offers a most inexplicable case of companionship in its habitual association with a certain species of Hermit-crab—the *Pagurus Prideauxii*. The Hermit or Soldier Crabs are well-known dwellers on the sea-beach, and ensconce themselves in the cast-off shells of whelks and other molluscs, for the purpose of protecting their soft bodies. On the shell which protects this veritable hermit, the cloak-anemone may almost certainly be found; and it is to be noted that only this species of crab, and the equally definite and single species of anemone, are the two beings which respectively form the association. The unvarying nature of the species is, in fact, as remarkable a feature in the case as the invariable nature of the companionship. And not only does the hermit-crab appear tacitly and simply to tolerate his living burden, with which, like Sindbad the Sailor and the Old Man of the Sea, he persistently crawls about, but he also appears to exhibit a certain care and affection for the anemone. He has been noticed to feed the anemone with his pincer-like claws; and when—as is the custom of these animals—the crab casts away his shell, to seek another and larger abode, he has been seen carefully to detach the helpless anemone from the old habitation, and to assist it in gaining a firm basis and support on the new shell. Another species of hermit similarly makes a companion of another kind of anemone; the latter subsisting on the food-particles furnished by its host. These details may pardonably suggest to us the idea that there may be, after all, much that is identical in the motives of even such lower forms as hermit-crabs, with the actions which we are accustomed, perhaps too exclusively, to regard as peculiar to ourselves.

The familiar little Pea-crabs, or *Pinnotheres*—so named from the small size of their bodies—present instances of a copartnership with salt-water mussels, the explanation of which is very hard indeed to find. Within the bodies of these mussels and of other molluscs, and within the folds of the structure which both lines and forms the shell, and which is appropriately named the 'mantle,' these little Pea-crabs appear to lodge in a perfectly natural and accustomed manner. As far as long-continued custom and habit are concerned, the Pea-crabs may well have become accustomed to their surroundings; for we find that Pliny of old, with other classical observers, was familiar with the fact of their unusual residence, and speculated on the causes which induced these animals to select their abodes. This old naturalist quaintly informs us that the mollusc being 'a clumsy animal without eyes,' opens its shell, and thereby allows other fishes to enter; and we are further informed that 'the Pinnothere (or Pea-crab), seeing his dwelling invaded by strangers, pinches his host, who immediately closes his shell, and kills, one after another, these presumptuous visitors, that he may eat them at his leisure.' Thus, the pea-crab is accredited at once with the virtue of efficient watchfulness and with the vice of jealousy; and so the case appears clear enough to this old naturalist, on the assumption that pea-crabs and molluscs are actuated by much the same motives as ourselves. The fact, however, of an active little body like the crab being allowed peacefully and naturally to dwell within the delicate, and usually irritable tissues of the well-known mussel, has as yet admitted of no satisfactory explanation at the hands of modern zoologists. Pea-crabs are also found living within those curious marine animals possessing bag-shaped bodies, and known as 'sea-squirts;' the crab dwelling within the breathing-chamber of its host. The author has noticed the crab to emerge from the mouth-opening of the sea-squirt to feed in an aquarium, in which its host was a tenant; the crab-guest beating a hasty retreat to its shelter on being alarmed. Pea-crabs measuring over half-an-inch in length may frequently be taken from mussels of not by any means large proportions.

The great insect-class exemplifies many remarkable associations, most of which, however, are examples of parasitism. For instance, a curious relationship subsists between ants and certain species of beetles. Indeed, some species of beetles which are totally blind, are nowhere to be found save in the nests of certain kinds of ants. These beetles are further known to be carefully tended by the ants, who at once attack any intruder into their nests, however nearly allied the latter may be to their blind friends. This instance of companionship is more mysterious than the well-known friendship that exists between ants and plant-lice, since the beetles do not, so far as observation has gone, furnish any secretion to, or otherwise benefit their hosts. One species of these blind beetles (*Claviger Duvalii*) is only found within the nests of a species of ant—the *Lasius niger*. Some ant-nests of this species may, however, be destitute of these beetle-visitors; and when the latter are artificially introduced into such guestless homes, the ants at once kill them. M. Lespès, who has given us these details, thinks that the latter fact may be accounted for by the supposition that some ant-colonies are more highly 'civilised' than others; but this explanation is more ingenious than probable or satisfactory.

Amongst Fishes, many examples of association with other fishes of widely different kinds, and for reasons not always apparent or explainable, are also to be found. The large, ungainly-looking fish possessing a very large head and wide mouth, frequently cast up on our shores after storms, and known as the Angler-fish or Fishing-frog (*Lophius piscatorius*), appears in many cases to give shelter, as a willing

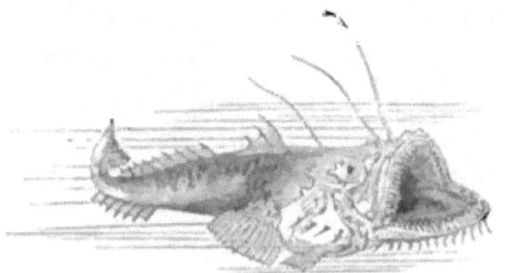

Fig. 56.—Angler (*Lophius piscatorius*).

or unwilling host, to a kind of eel, which lives within its capacious gill-chambers. The eel-guest doubtless subsists on the food-particles which may find access to its abode, from the equally capacious mouth. The well-known Pilot-fish has received its name from its supposed habit of piloting

Fig. 57.—Pilot-fish.

sharks towards their prey; whilst, as was believed by the ancients, it also warned the sea-monster against dangers of all kinds. Of the mere fact of the companionship between sharks and the pilot-fish, there can be no doubt; but it seems to be doubtful if the attendance is of the disinterested kind just alluded to ; as the contents of the stomach in the pilot-fish, we are told, generally consist of food which it has picked up for itself. It is therefore not a mere parasite, but may probably follow the shark from the expectation that its chances of picking up food are greatest in the neighbourhood of so powerful a caterer.

The Remora, or Sucking-fish, in virtue of possessing a peculiar sucker on the top of its head, forms associations with other fishes, probably as an aid to locomotion. Fixed to the body of another fish, this clinging companion is saved all further trouble of movement on its own account, and roams wheresoever its foster-friend may list. The ancients, it is curious to note, thoroughly believed in the powers of the remora to detain, by an exercise of immense or supernatural strength, any objects to which it might attach itself. Antony's ship, at the battle of Actium, was reported to have been held fast by a remora, and the vessel of Caligula was alleged to have been similarly arrested. The fish itself attains the length of twelve or thirteen inches, and somewhat resembles a herring in its general shape.

In the class of Birds, many notable examples of curious likes and dislikes of personal kind, if we may so style them, may be found. For whilst in some cases the friendly companionships are very evident, so no less are examples of aversions and dislikes. The cuckoos thus present us with curious instances of semi-parasitic habits, in their invasion of the nests of other birds for the purpose of depositing their eggs ; and the association between the birds known as Ox-peckers (*Buphaga*) and cattle, is no less curious in its details, even if we consider that the reasons for the companionship are of very evident kind. The ox-peckers form a group of Perching Birds, inhabiting Africa ; a familiar species being the Common Ox-pecker (*Buphaga Africana*); and their popular name, together with the designation—not applied to birds alone—of Beef-eaters, has been given to them from their habits of following herds of cattle in great numbers, and of perching on the backs of their bovine neighbours, for the purpose of extracting the larvæ or caterpillar-forms of the troublesome bot-flies. The eggs of these flies being deposited in the back of the ox, and usually in a part which the animal is unable to reach with his tongue, give rise to a troublesome swelling, known as 'worble,' within which the young insects undergo part of their development. The ox-pecker alighting on the back of the ox, soon contrives, by aid of his powerful and peculiarly shaped bill, to extract the larvæ—an operation seemingly conducted with gentleness and skill, and apparently relished, as a relief from pain, by the subject of the operation; the oxen evincing no uneasiness or objection, consequent on the attentions of these birds. In like manner, starlings in our own country befriend sheep by ridding them of troublesome larvæ. In short, it would be difficult to find more typical cases of true co-operation for the purposes of mutual benefit, than those before us.

As an example of aversions on the part of very closely related birds, may be cited the case of the swift and chimney-swallow; these birds being rarely, if ever, seen to associate together; while the more positive fact of their aversion is exemplified in the instance, familiar to all ornithologists, that when these two genera of birds take up their abode in one street, the swallows will select one side, whilst the swifts retain the other. This conservatism in nest-building extends to their more active habits; for, when in flight, the two genera, so much alike in appearance and in their selection of food, appear to preserve the same air of restraint and non-companionship. And the consideration of the present case is rendered all the more puzzling in its aspect, from our knowledge of the fact, that the house martin and chimney-swallow, in their earlier years at least, are close companions and friends.

Amongst higher animals than birds, instances of the preceding traits of character are by no means wanting. Thus, as far as unwonted familiarity is concerned, the expression 'cat-and-dog life' is not always synonymous with hatred and discord; but is sometimes, on the contrary, indicative of the closest and most friendly attachment. A raven and a cat have become lifelong friends; and rats and dogs, and cats and mice, have been known to lay aside their inherited differences, and to fraternise in the most amicable manner. Occasionally we may meet with examples of aversion amongst quadrupeds which are not readily explained; as, for instance, the commonly observed fact, that horses and cows grazing in the same field rarely fraternise; whilst cows and sheep appear to be less conservative in their habits and associations.

In zoological collections, companionships of unusual kind are not unfrequently formed; and, although such traits of character are less surprising amongst animals of high intelligence, such as monkeys, instances are not wanting to shew that, as in some human friendships, the 'contracting parties' may be of very dissimilar kind. A large and powerful monkey has thus been known to staunchly befriend and protect a weak and insignificant brother, of different species, from the attacks of the other occupants of the cage, and also to reserve the delicacies, which his superior strength could secure, for his less agile companion.

Some very curious, but at the same time uncertain cases of animal companionship, are constituted by the association which is alleged to exist between those largest of living reptiles, the crocodiles, and certain kinds of small birds. It has undoubtedly been observed that some small birds continually hover around the haunts of these reptiles. The older naturalists firmly believed that the birds befriended their reptilian neighbours, and gave them due notice of the approach of enemies; but it is possible that this assertion has originated rather as a theoretical explanation of the companionship—if such association really exists—than as an observation founded on fact. The origin of such a mode of explaining animal companionships has already been illustrated by Pliny's account of the nature of the association between the Pea-crabs and their hosts the mussels.

Sometimes, however, a case of apparent association may be disproved by closer scientific scrutiny. A remarkable worm-like organism had, for example, been long known to occur in invariable association with certain cuttle-fishes. These 'worms' were figured and described by various naturalists under the name *Hectocotyli*, and every one appeared to be satisfied of their parasitic nature and life. But to the astonishment of the zoological world, more careful observation afterwards shewed that the supposed parasite was in reality one of the modified arms or tentacles of such cuttle-fishes; the altered appendage having obvious bearings on the development of these animals.

ANIMAL DISGUISES AND TRANSFORMATIONS.

MOST people are aware, as a piece of commonplace knowledge, that many animals, before arriving at their mature or adult state, undergo a series of changes in form, of a more or less complete character. To such a series of changes the naturalist applies the term 'metamorphosis;' and the study of the disguises which an animal may in this way successively assume, forms one of the most interesting and fascinating subjects that can attract the notice of the observer.

The great insect-class presents us with the most familiar examples of these changes, and the butterflies and moths exemplify metamorphosis in its most typical aspect. Thus we know that from the egg of the butterfly, deposited by the short-lived parent upon the leaves of plants, a crawling grub-like creature is first developed. This form we name the 'larva' or 'caterpillar' (fig. 58, *a*); and if we might fail to recognise its relationship to the bright denizen of the air as far as outward appearance is concerned, we might also be at a loss to reconcile its internal structure with that of the perfect butterfly. Thus the latter is winged; possesses a mouth and digestive system, adapted for the reception and assimilation of flower-juices; and wholly differs in structure and habits from its worm-like progeny. The caterpillar is provided with a mouth furnished with jaws, and adapted for biting or mastication; its digestive system presents a type differing widely from that of the perfect form; and its crawling, terrestrial habits appear in strong contrast to the light and ethereal movements of its parent.

The life of this larva may be accurately described as one devoted solely to its nourishment. Its entire existence, whilst in the caterpillar state, is one long process of continuous eating and devouring. By means of its jaws it

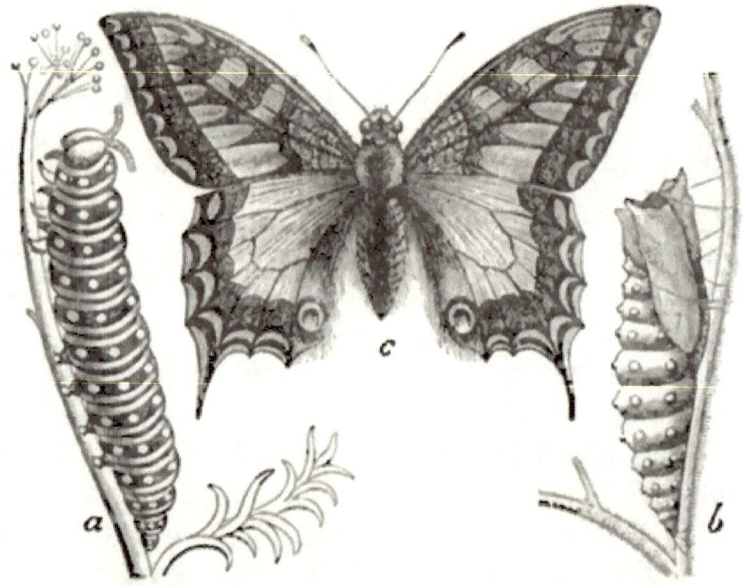

Fig. 58.—Development of a Butterfly:
Metamorphosis of the Swallow-tailed Butterfly (*Papilio Machaon*): *a*, Larva or Caterpillar; *b*, Pupa; *c*, Imago, or perfect form.

nips and destroys the young leaves of plants, much to the gardener's annoyance; and so rapidly does its body increase in size, that the first skin with which its body is provided soon cracks and bursts like a tight-fitting coat, and a process of moulting ensues. As the result of this process the larva emerges, clad in a new skin, adapted to the increased size of its body. This second skin may similarly become inadequate to accommodate its ever-increasing growth, and a second process of moulting produces in turn a new investment. In this way the caterpillar may change its coat many times—twenty-one moultings have been counted in the development of the May-flies—and on arriving at the close of its larval stage of existence, may present a very great increase in size, as compared with the dimensions it presented at the beginning of its life.

But, sooner or later, the caterpillar appears to sicken, and to become quiescent. Its former state of activity is exchanged for one of lethargy, from which it awakes to begin an operation of a novel and different nature from that in which it has been previously engaged. It begins to spin a delicate silky thread by means of a special apparatus, situated in the head, and which consists of silk-glands, and of an organ named the 'spinneret.' Within the silken case or 'cocoon' which it thus constructs with the thread of the spinneret, the caterpillar-body is soon enclosed; the first stage of its existence comes to an end; and the second or cocoon stage, marked by outward quiescence and apparent rest, becomes known to us as that of the 'pupa,' 'chrysalis,' or 'nymph' (fig. 58, b).

Although outwardly still, and although all the former activity appears to have been exchanged for a state of dull repose, changes of active kind, and of marvellous extent, are meanwhile proceeding within the cocoon or pupa-case. The elements of the caterpillar's form are being gradually disintegrated or broken down, and built up anew in the form and image of the adult butterfly. Old textures and garments are being exchanged for new ones; particle by particle the outward and inward structures of the larva are being replaced by others proper to the mature being; and in due course, and after a longer or shorter period, the cocoon is ruptured, and the perfect form emerges—a bright and beautiful creature, furnished with wings and active senses, and rejoicing in the exercise of its new-born functions amid the sunlight and the flowers.

Such is an outline of the familiar process by which the larva or caterpillar of the butterfly becomes transformed or developed, to form the 'imago' or perfect and adult form. And if we review the stages exemplified in the process, we shall be able to detect in each an obvious harmony and correspondence both with the preceding and with the succeeding stage. Thus we find that the life of the perfect and mature insect is at the best of a comparatively short and transient nature, and its energies are directed chiefly to reproduction—to the deposition of eggs, from which new individuals will, in due course, be produced. The larval stage, on the contrary, is devoted to nutrition; to the laying up, as it were, of a store of nourishment, sufficient to last throughout the lifetime of the being, and to sustain it whilst its adult functions are being performed.

Indeed, the entire lifetime of the higher insect may be divided into, or comprised within, two distinct periods. The first of these latter is the nutritive period, represented by the caterpillar-state, when the nutrition of the body is mainly provided for; and the second period, no less defined than the first, is included in the life of the perfect form, which is devoted to reproducing the species. This last we might therefore term the reproductive period of insect-life.

All insects, however, do not exemplify 'metamorphosis' in so perfect a manner as does the butterfly. The beetles, flies, bees, &c., and many other insects, undergo a process of metamorphosis essentially resembling that of the butterfly; the characteristic feature of this form of development being that whilst the caterpillar stage is passed in activity, the pupa or chrysalis is quiescent; and from this resting-pupa the active, winged insect comes forth. The dragon-flies, crickets, grasshoppers, and their allies, undergo, on the other hand, a less perfect series of changes than the foregoing insects. The young grasshopper, on leaving the egg (fig. 59, *a*), bears firstly a close resemblance to the perfect insect. It is, further, not of worm-like conformation, and in these two points differs from the larva of the other forms. Then, thirdly, it does not enclose

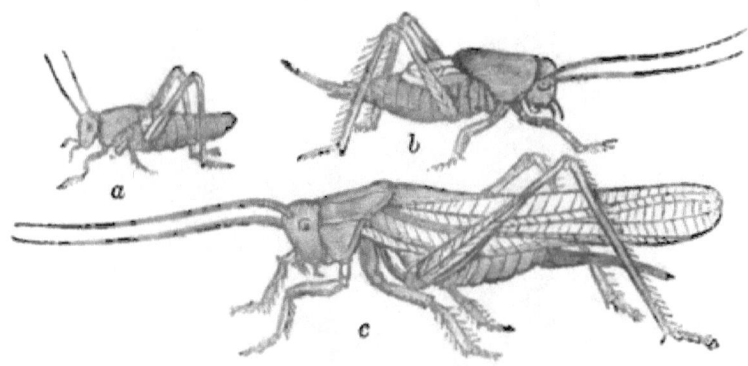

Fig. 59.—Metamorphosis of Grasshopper:

Incomplete metamorphosis of Grasshopper (*Gryllus viridissimus*): *a*, larva; *b*, pupa, in which the wings are beginning to appear as lobes on the hinder portion of the chest; *c*, imago or perfect insect, characterised by the possession of fully-developed wings.

itself in a cocoon-case, but passes its chrysalis stage in a free and active condition. In this respect it again differs from the butterfly chrysalis; and its perfect form is attained simply by the development of the wings. So that, in reality, the chief difference between the young and the perfect form of the grasshopper consists in the non-development in the former of the wings, which are thus characteristic of the adult form.

The Dragon-flies illustrate an essentially similar kind of metamorphosis, but also exemplify differences in the details of their development. The young of the dragon-fly are active creatures, inhabiting the water of pools; the eggs from which they are produced having been deposited by the parent in bunches on the leaves of water-plants. The larvæ, *b*, are of brownish colour, and possess six legs, and a peculiar apparatus of jaws, consisting of a pair of nippers attached to a movable, rod-like stem. This apparatus can be folded upon the head, when it gives to the larva the appearance of being

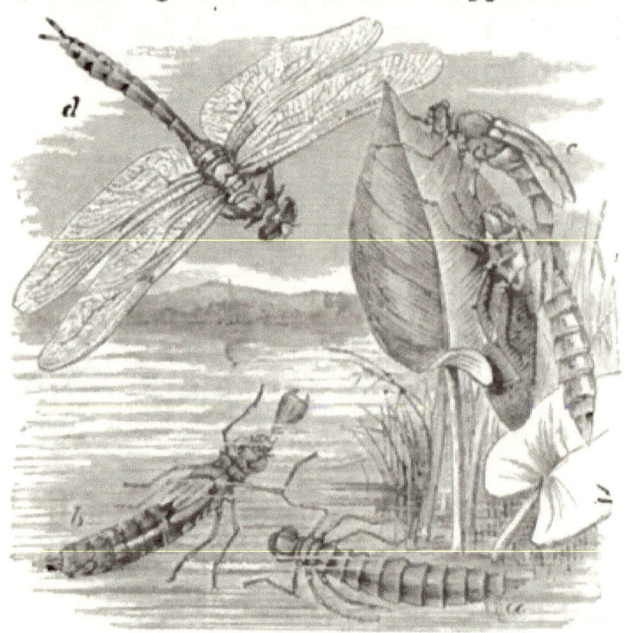

Fig. 60.—Metamorphosis of Dragon-fly:

a, larva; *b*, pupa; *c*, perfect insect issuing from pupa-case; *d*, perfect insect, with wings fully developed.

masked, and hence the name of 'mask' which has been applied to this structure. But on the approach of some unwary insect, the jaws can be rapidly extended to seize the unfortunate victim, and convey it to the mouth of its captor. The dragon-fly's young are thus purely aquatic in habits, and propel themselves along by ejecting water, which has been used in breathing, from the posterior extremity of the body.

Having arrived at the close of its chrysalis-stage of

development—the chrysalis differing from the larva simply in its greater size, and in the development of the wings and perfect body within the pupa-skin—the insect at length fixes its body to some water-plant. The pupa-skin next splits along the back, and the mature, winged insect slowly emerges therefrom. The crumpled wings soon dry, and acquire their normal consistence; and the dragon-fly, freed from the trammels of a mundane existence, mounts into the air, and 'revels in the freedom of luxury and light.' Tennyson has aptly described this change in his lines:

> To-day I saw the dragon-fly
> Come from the wells where he did lie.
>
> An inner impulse rent the veil
> Of his old husk: from head to tail
> Came out clear plates of sapphire mail.
>
> He dried his wings: like gauze they grew:
> Thro' crofts and pastures wet with dew
> A living flash of light he flew.

In these latter instances, as in the case of the butterfly, the nutrition of the insects has been proceeding during the earlier stages of life, and has been fitting them for entering upon the final part of their existence, which may extend for a longer or shorter period, but which is mainly devoted to the continuation of the species. The time occupied in the development of insects varies greatly in different groups. Cold and damp appear to delay this process. The chrysalis of a butterfly has been kept for two years in an ice-house, without undergoing development; whilst on removal to a warm place it became transformed into the winged insect. The Cockchafer occupies three years in its development, the duration of life in its perfect state being probably only a single year.

In the frogs, toads, newts, and their allies, as representing the higher vertebrate animals, we find well-known and interesting examples of changes in development. The larval frog appears before us as the familiar tadpole, which breathes at first by outside gills, and then by gills developed within the body. Its form and breathing are thus at first fishlike, and it swims by aid of its elongated tail, which is provided with a delicate fin. The tadpole further feeds upon water-plants, which it nibbles by means of the horny jaws with

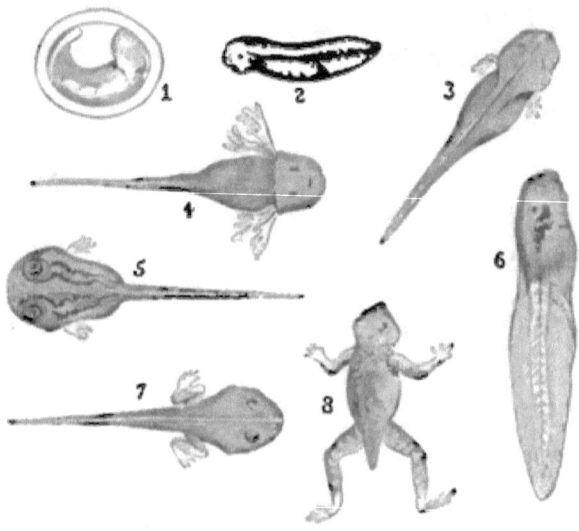

Fig. 61.—Development of the Frog:
Successive stages—in the order of the numbers—from the egg almost to the perfect form.

which it is provided; and it also possesses suckers on the under-surface of the head, and attaches itself to fixed objects by aid of these organs. The hind-limbs first appear as little buds from the posterior portion of the body, and the fore-limbs soon follow. Then the tail begins to shrivel up, and to become rudimentary; lungs are meanwhile being developed; the gills next disappear, and, finally, the frog leaves the water, and becomes for the remainder of its life an air-breathing and terrestrial animal.

The foregoing cases present us with a few examples of the 'disguises' which animal forms may assume during their development from the egg; and we may at the present stage very briefly inquire as to the nature or reason, if any may be found or suggested, for the occurrence of such phenomena. Broadly speaking, the young of the insect undergoes the greater part of its development outside the egg and parent-body; and it thus differs, in one sense, only in the mode and place of its development, from the progeny of other and higher animals. It was long ago held that the most perfect examples of metamorphosis occurred in those animals the eggs of which contained little or no nourishment for the sustenance of the developing young; the offspring in such a case obtaining nourishment independently of the parent-body, and as it grew.

These explanations, however, deal rather with the results than with the origin of metamorphosis. Why, in one case it should be so well marked, and, in other cases not occur at all, form considerations which have long puzzled the naturalist. It has been maintained by certain zoologists that the changes which any animal may in the course of its development undergo, illustrate its relationship with other animals, from which it may have descended, or with which it may possess relations of a genealogical kind. Metamorphosis has thus been pressed into the service of the theories of evolution, which, as most people are aware, hold that all animals have descended from previously existing animals, or have been evolved from their predecessors, by or through various processes. Thus we find Mr Darwin maintaining that 'the embryonal (or young) state of each species reproduces more or less completely the form and structure of its less-modified progenitors;' and according to this view, we would therefore see in the young crab, with its larval tail, a transient representation of the lobster-like progenitor from which the short-tailed crab-race was in past times developed and evolved. And with regard to insects, it is held that external or outside forces and conditions, acting upon the young or larval state, have had much to do, in the past, as well as in the present, with producing the differences between the various groups of insects; the metamorphosis of which presents us with a panoramic view of their origin and modifications. Whether or not these conclusions are true and good ones, time and the progress of research alone can tell; but the importance and interest of such a study as that which forms the subject of these remarks, cannot be lessened by any theoretical considerations which become interwoven with it.

In connection with the frogs and their neighbours it is interesting to note some facts which have been recently brought to light regarding the curious changes of structure exhibited by the quaint-looking amphibians found in Mexico, and named *Axolotls*. These animals (fig. 62) are somewhat lizard-like in general appearance, and possess compressed tails adapted for swimming; whilst from a cleft in each side of the neck three beautiful tufts of gills may be seen to protrude. The axolotl thus breathes by these outside gills, but the animal also possesses simple lungs; so that it is truly an 'amphibian,' and is adapted for living both in water and on land. Until 1867, the axolotls were thought to represent a distinct species of animals; but in that year the fact was made known that they may lose their outside gills, apparently in a perfectly natural manner, and may assume the exact form of a purely land salamander inhabiting North America, and long known under the name of *Amblystoma* (fig. 63). The Amblystoma, like the frog, breathes by lungs alone in its full-grown state, although it resembles the latter animal in possessing gills in its young condition. The startling fact was thus brought to light, that an animal long thought to be

Fig. 62.—Axolotl.

a full-grown and perfect creature, and which had given every evidence of being an adult form, could naturally assume the higher form of a purely air-breathing neighbour. These facts

Fig. 63.—Amblystoma.

were brought to light by the observation of axolotls which were kept in confinement; about thirty of these animals belonging to the Jardin des Plantes at Paris, having in 1867 shed or cast off their gills, and appeared as permanent land-dwelling beings. A French naturalist, wishing to see if the artificial removal of the gills would have any effect upon the axolotl's change of nature, cut off the gills of one specimen; but the animal, instead of being encouraged to a new mode of life, simply developed new gills to replace the lost structures, and remained persistently as the well-known axolotl.

A German lady, Fräulein von Chauvin—whose name deserves to be held in remembrance by zoologists—has, however, recently shewn that by patience and perseverance in the treatment of the axolotl, and in the regulation of its mode of life, it may be made to assume the form of the *Amblystoma*. The axolotls experimented upon by this lady, were first induced to adapt themselves to a life on land, by the water in which they were contained being made shallower. Some specimens which were ultimately taken wholly out of water did not thrive well, and great difficulty was experienced in feeding them; Fräulein von Chauvin ingeniously thrusting a worm head-first into the patient's mouth, and causing it to wriggle downwards into the throat, so that the refractory axolotl was obliged at last to swallow the morsel. After a life of fifty days on land, three out of the five animals selected for experiment died, exhibiting, however, at the time of their death, a marked decrease in the size of the gills. The two survivors, however, flourished on land, and the tail-fin and gills literally grew 'small by degrees, and beautifully less,' with the result that when the axolotls were restored to their former element, they made all haste to get back to *terra firma*. As time passed, not only did the gills shrivel up and disappear, but the clefts through which they were emitted also closed up. The tail lost its fin, and became rounded like the tail of Amblystoma; whilst the skin—soft and moist in the axolotl—became of denser texture, and developed the black and yellow colour of the land animals. The axolotls, thus metamorphosed into Amblystomas, evidently found their appetite anew with their change of structure, and ultimately required no coaxing to cause them to take their natural food.

These curious facts exhibit to us the means whereby Nature may, in the past, have produced the races of these air-breathing animals, by the modification of water-dwelling animals. And no less instructive are the details of the disguises under which, on the other hand, a truly land-living and air-breathing animal may be made to adapt itself to a water existence. Like the frogs and all its other neighbours, the Black Land-salamander of the Alps begins life as a tadpole possessing gills; only, having no access to water on the mountains, the gills are cast off before the young leave the parent-body, so that after birth these animals breathe by lungs alone. Fräulein von Chauvin having obtained two young salamanders before they had been hatched, and when they still possessed gills, placed these young gill-possessing forms in water. One soon succumbed to the change of habitation, but the other, after four days' life in the water, cast off its first gills, and developed a second set, and also a tail-fin for swimming. For fifteen weeks this young animal—which in the course of its natural development should have been running about on land as a true air-breather—existed below water, apparently making the most of its existence. But after that period, the gills were cast off, lungs were developed, and the erring salamander returned to the ordinary life and ways of its kind. In this case Nature was tempted to reverse the usual order of development, and to permit an animal to remain for a time in a state inferior to that of its adult and ordinary existence. We are again taught an important lesson, not only regarding the power most animals possess of accommodating themselves to outward circumstances, but also concerning the important and permanent effects these circumstances may induce in the life and structure of living beings.

Sometimes, however, we meet with cases in which the transformations which an animal undergoes, lead not to a higher or superior state, but terminate in a condition lower

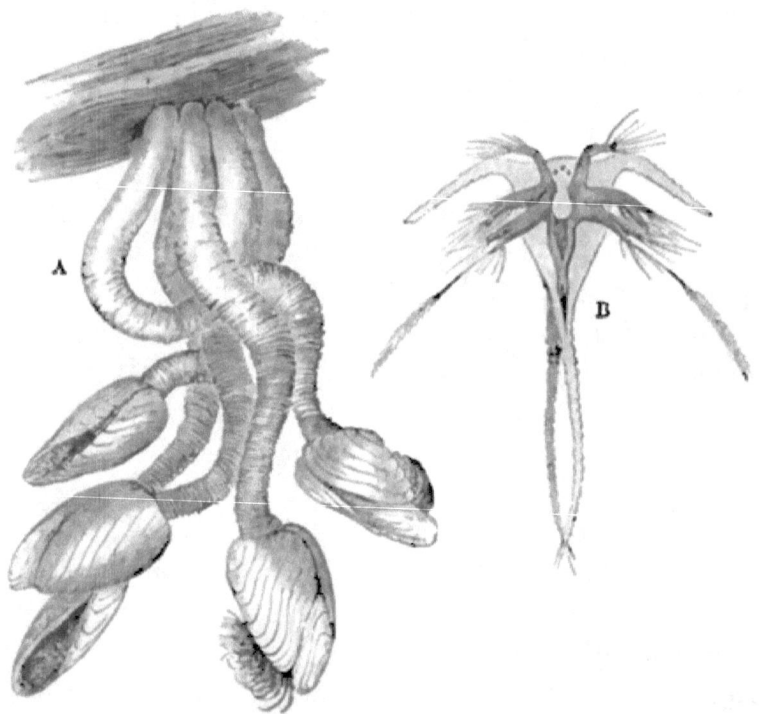

Fig. 64.—Development of Barnacle:
A, group of Barnacles; B, young or Nauplius of Barnacle.

than that exhibited by its young or larval period. The familiar Barnacles (fig. 64, A) and Acorn Shells which incrust floating timber and rocks, begin their existence and leave the egg in the form of little creatures (fig. 64, B)—each known as a *Nauplius*—which swim actively through the water.

The body of each of these larvæ is somewhat oval in shape, and is covered by a buckler-like structure; whilst, like the Cyclops of old, it possesses a single eye. An elongated 'tail' is developed, together with three pairs of swimming-feet; and altogether the young animal is as different from the fixed barnacle as could well be imagined. This larval form, like the caterpillar of the insect, moults frequently, and then the nauplius becomes a pupa or chrysalis. In the latter stage, the body is enclosed in a 'shell' consisting of two pieces; whilst the first pair of limbs or feet have become converted into large 'feelers,' destined to play an important part in the future history of the animal. The final stages in this curious development are ushered in by the casting away of the two remaining pairs of feet of the larva, and by the development of six pairs of strong jointed feet on the under part of the body of the pupa, which at this stage possesses two compound eyes. Ultimately this wandering pupa attaches itself by the large feelers to some piece of drift-wood; these organs being glued to the fixed object by a special 'cement.' The eyes are next cast off; the shell of the adult barnacle becomes developed; and the six pairs of feet are transformed into the tentacle-like organs (fig. 64, A), which in the living barnacle are seen to be continually sweeping in and out of the shell. Thus we note in this case, that a free, active being, provided with eyes and other organs, becomes developed into a stay-at-home, eyeless barnacle; the process of development thus illustrating what is known as 'retrogression,' or physiological backsliding.

In certain relations of the barnacles, we may meet with still greater differences between the young and the adult animal; the former appearing to possess all the advantages of perfect structure, and the latter appearing as a forlorn and destitute creature in comparison with its earlier stages. Attached to the bodies of certain kinds of crabs, little sac-like animals may be found. From these animal-sacs (fig. 65, A) roots are given off, and these roots penetrate into the body of the crab, and serve to attach their possessors firmly to the latter animal. Each of these little bags is appropriately named a *Sacculina*, and as we note the sacculina attached to

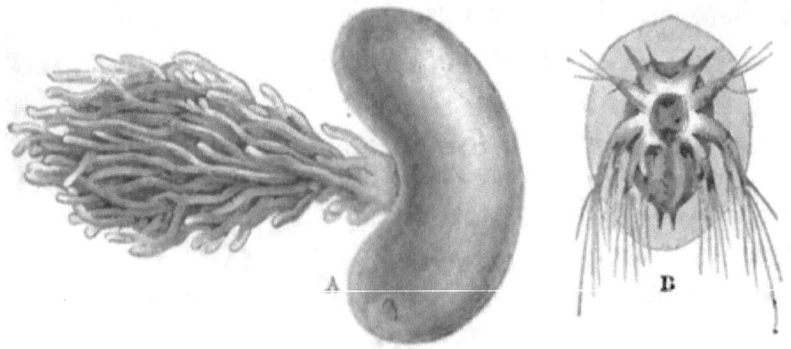

Fig. 65.—Development of Sacculina:
A, Sacculina; B, young Sacculina.

the crab, we might well think it to represent simply some curious and unusual outgrowth from the body of that animal. If we slit open the sac, we find it to contain eggs, but we can gain thus no information regarding the exact nature of the structure or being, whatever we may term it.

Let us, however, watch the development of one of the eggs of the Sacculina-body, and we shall obtain materials for a curious study in animal transformations. From each egg there will issue a little oval body (fig. 65, B), covered with a back-shell, and provided with four feet, to which long bristles are attached, and also possessing two long feelers. We thus see that the sacculina-egg has developed into a being resembling the young barnacle; and like the latter form, the young sacculina swims merrily through the sea. Soon the shell of the back becomes folded, and develops two pieces, like the larva of the barnacle; and as in the latter, six pairs of additional feet appear. The young sacculina and the young barnacle might in fact be regarded as identical, since at this stage no difference is perceptible between them.

The feelers next become greatly developed, and grow into the form of branched root-like organs. The young sacculina then seeks out some crab-host, and attaches itself thereto by means of its root-like feelers. The shell and other organs drop off, and the formerly active and free-swimming being becomes thus transformed into the inert bag-like sacculina, the eggs of which will each repeat the curious cycle of development through which their progenitor itself has passed.

Such a study, besides revealing to us some very astonishing phases of animal life, shews us how animals, differing

Fig. 66.—Pentacrinus Caput-medusæ.

widely in their full-grown state, may be proved to be in reality closely related through their development. The Sacculina is thus seen to belong to the class Crustacea, to which not only the barnacles, but the crabs, lobsters, and their neighbours belong; similarity in development being reasonably regarded as proving the true identity and relationship of animals.

A concluding instance of curious features in animal transformations is presented by the development of the beautiful little star-fish known as the Rosy Feather-star— the *Antedon rosaceus* of the naturalist. In 1823, Mr J. V. Thompson discovered in the Cove of Cork a little stalked creature, which he named *Pentacrinus Europæus*, and which was thought to be a new species of Crinoids or lily-stars (fig. 66), each of which may be compared to a star-fish set on a stalk. Imagine the astonishment of the discoverer, when in 1836 he for the first time saw the star-fish head drop off from its stalk; the detached star-fish appearing as the well-known rosy feather-star, of whose development the scientific world was previously ignorant. But the feather-star is at the same time a true crinoid, and not related, in a close fashion at least, to the star-fishes. It simply differs from other crinoids in being stalked during the earlier portion of its existence only.

ANIMAL ARMOURIES.

THERE can hardly be any greater diversity observed in the animal series than that exemplified in the various means whereby animals are enabled to assume an offensive or defensive aspect. From the lowest to the highest grades of animal life—excepting perhaps man himself—we find ample provision made for the exigencies of animal existence, in so far as these exigencies demand the use of some apparatus which gives its possessors some advantage or other in the 'struggle for existence.' Undoubtedly, in his superior intellectual organisation, which enables man even in his rudest state to avail himself of almost every feature in his surroundings for advantage and defence, the human subject has been endowed above all other forms; and he therefore compensates himself by varied arts and stratagems for the want of the more rigid and natural appliances of lower forms. But if it be true that art is most to be admired when it closely imitates nature, then the policy of man in his imitation, conscious or unconscious, of the many offensive arts of his humbler neighbours, must claim from us a fair share of favourable criticism.

Thus, it is a striking fact, that very many human means of defence or offence find their prototypes, or at least strangely analogous features, in the extensive armoury of the animal world at large. The lasso may be found imitated in the apparatus whereby such a simple form as the *Hydra*, or tiny Fresh-water Polype (fig. 67), secures its prey. Or, when human sharp-shooters think to conceal their whereabouts most effectually from the foes they purpose to annoy, and clothe themselves in garments of neutral tint the hue of which shall most nearly resemble that of the objects amidst which they are located, this principle of imitation of natural objects again finds a strict parallelism in the animal world. For it is a familiar fact to all observers of nature, that the colour of most animals resembles more or less that of their natural surroundings. The colour of the sand-grouse, for instance, and other species of grouse, of partridges and other birds inhabiting heaths, or of flounders and other fishes inhabiting the sand, strictly approximates in character to that of their dwelling-places, and serves to conceal and protect such beings. The woodcock, for example, is not easily distinguished on the ground. Butler in his *Hudibras* says that

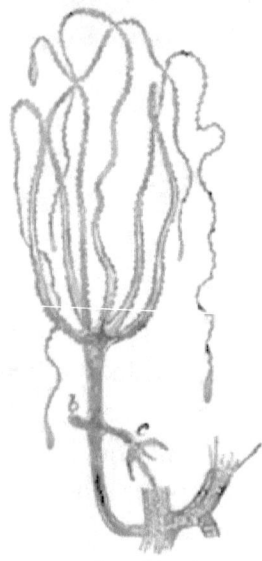

Fig. 67.—The long-armed Hydra: With a young bud at *b*, and more advanced bud at *c*.

> Fools are known by looking wise,
> As men find woodcocks by their eyes—

the remark of the satirist having reference to the bright black eyes of these birds. And when we further discover that, in not a few cases, this principle of similarity to their surroundings is carried in some animals—such as the Leaf-insects and Walking-stick Insects—to the extent of close and actual *mimicry*, our surprise is increased.

Or lastly, when we find, as in the latest phase of modern warfare, that the concealed torpedo is used as a subtle and powerful means for effecting the destruction of whole fleets, the fact cannot but call to mind the electrical apparatus of some fishes—and notably that of the *Torpedo* or Electric Ray—which exists as a natural means of defence, the powers of which, few, if any, of their less-favoured neighbours care to test or provoke.

Whilst the consideration of the more prominent and typical means of defence in animals may very reasonably occupy our brief attention, a few words on the subject of Mimicry in the animal series may also prove interesting, more especially as this form of protection, through imitation of their surroundings, forms a simple yet effective means of defence to many beings. We have already referred to the readily perceived and very general correspondence in colour seen throughout the animal world between animals and their abodes; and of the more general aspects of this condition nothing further need be said. The more special and striking developments of mimetic resemblances are found in cases in which not merely the general colour of their environments is imitated, but where resemblances of a close, and sometimes of a very extraordinary kind, to other animals, to plants, or even to inorganic objects, are to be noted. In the leaf-insects (fig. 68), which are included in the same order as locusts, crickets, &c., for example, the wings are not only coloured to resemble leaves, but their structure imitates in the most exact manner the appearance of the veins of the leaf. Nor does the principle of imitation end with this sufficiently remarkable effect. In some leaf-insects the colours of the leaf-like wings actually change with the season of the year; as if in the most perfect sympathy and harmony with the alteration of colours in the actual leaves. And the mimicry becomes of still more perfect kind, to our thinking, when we find that the wings of the leaf-insect

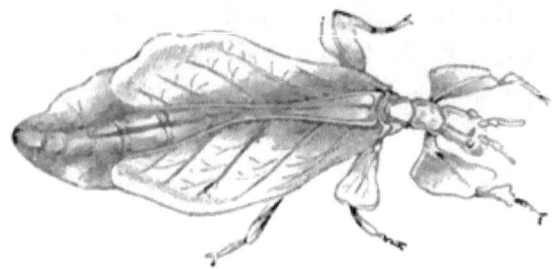

Fig. 68.—Leaf-Insect.

exhibit even the characteristic markings we are familiar with in leaves as produced by the attacks of minute insects and of parasitic plants; Nature thus imitating not merely the natural structure of the leaf, but the very imperfections and diseases to which the leaf is subject. It has been suggested that little leaf-eating parasites may be themselves deceived by the mimicry of their larger neighbours, and may actually eat into the wings of the latter, and thus produce the eroded appearance. But even if this latter view be correct, it only makes out a stronger case for the perfect representation of the leaves in the wings of the insect. Mr Wallace has given us a very typical example of another such case of the imitation not only of leaves, but of the natural parasites of leaves, in a butterfly, the wings of which, on their under-surfaces, resemble leaves; whilst the imitations of decay of leaves and of the fungi that appear thereon, are so close, that, as Mr Wallace remarks, 'it is impossible to avoid thinking at first sight that the butterflies themselves have been attacked by real fungi.'

The Walking-stick or Spectre Insects (fig. 69), as they are called, in their turn imitate, in the skeleton-like structure of their bodies, the appearance of dried twigs; and it is a singular fact that even in their awkward, ungainly manner of walking, the resemblance to the movements of twigs disturbed by the wind is clearly perceptible; the mimicry being rendered more realistic through this latter phase. Then, also,

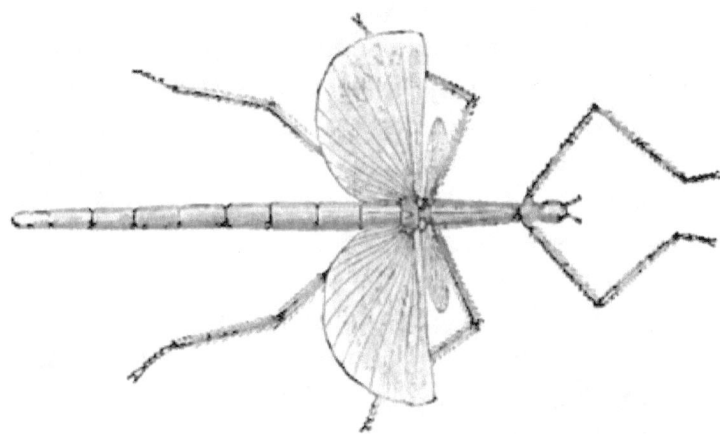

Fig. 69.—Walking-stick Insect.

we find certain harmless groups of moths imitating closely the outward appearance of species of stinging bees and hornets. Another remarkable case of mimicry is the well-known instance of some inodorous South American butterflies, which perfectly reproduce the external appearance of other butterflies, the latter emitting a most offensive odour. The reason assigned for this latter phase of mimicry is the very feasible one, that the inodorous forms are protected from the attacks of birds by their resemblance to their strong-smelling neighbours. As a last instance of this curious phase of animal organisation, we may note the example furnished by those curious little fishes, the *Hippocampi*, or Sea-horses. The bodies of these fishes become covered with long streamers of certain kinds of sea-weed; so that when these fishes rest amidst the sea-weed-covered nooks of their marine grottoes, the presence of their streamers serves to render detection by their enemies no easy matter.

Referring to the explanation, if such can be afforded, of these mimetic resemblances, there can be little doubt that, viewed as to its ultimate use and purpose, the condition of mimicry serves in the most effective manner as a means of defence and protection to the animals so endowed. The resemblance of the colours of birds to that of their habitat, presents an obvious instance of this purpose; as also does the more complicated example of the imitation by scentless butterflies, of their odorous neighbours. But as regards the exact means whereby the condition of mimicry is attained and perfected, or concerning the exact causes of its assumption and development, natural history science in its practical aspect remains silent; although the bolder march of theory and speculation may indeed lead us for a little way towards the solution of the problem. At anyrate, there can be no difficulty to our clearly appreciating the workings of a great law of purpose and design in the production of mimicry, as serving to protect the weak and less powerful against stronger and better-provided animals.

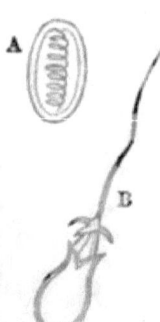

Fig. 70.—Thread-cells:
A, thread-cell of Hydra (at rest); B, with thread everted.

Turning now to some lower forms of animal life, we find in such forms as the *Hydræ*, or common Fresh-water Polypes, the Zoophytes, Sea-anemones, Jelly-fishes, and allied forms, excellent examples of very specific means of defence and offence in animals. Within the tissues of the bodies of these organisms, when these tissues are microscopically examined, numerous little sacs or cells, varying in size and form, may be observed. To these cells the appropriate name of 'thread-cells' (fig. 70) has been given. When their structure is investigated, each little cell is seen to possess an elastic wall of double nature; the inner layer of the wall being strong, whilst the outer one is of thinner and more delicate texture. The upper or open extremity of the inner layer of the sac is prolonged to form a kind of sheath, which protects and gives origin to a thread-like filament, from the presence of which, indeed, these cells derive their name. This 'thread,' in the ordinary condition of the cell, is coiled up within the interior of the sac, and around its own sheath; and in many cases both thread and sheath may be discerned to be provided with minute spines or hooks. The cell itself, in addition, contains a fluid, amidst which the thread is submerged.

Such is the essential structure of a thread-cell in its normal state of what we may term repose. When such a structure, however, is pressed or irritated in any way, the cell ruptures or bursts, the contained fluid escapes, and the thread and its sheath are quickly protruded or thrown out from the opening in the cell. If now the thread and fluid are observed to come in contact with any body of appropriate and assailable kind, such a body will exhibit certain symptoms which will indicate to us the probable nature of these curious cells. Thus, when the tentacles or feelers of the Sea-anemone, or of any of the Zoophytes (fig. 67), come in contact with a minute or susceptible organism adapted for food, the prey is first observed to struggle to escape from the entwining filaments which encircle its body. Soon, however, its active exertions cease, and the victim appears paralysed and incapable of helping itself, or of struggling longer with its captor. The thread-cells, in other words, have been discharging their miniature darts or 'threads' into the body attacked; the fluid—in all probability, of acrid or poisonous nature—has been poisoning the tissues of the struggling organism; and the observation has revealed to us that the functions of the cells are undoubtedly analogous to those of the serpent's fangs and poison-gland, in that they serve to paralyse and kill the prey.

As might naturally be supposed, the power of the thread-cells varies in different species and groups of the animals that possess them; but there are some forms of Cœlenterate animals—for thus the Hydræ, Sea-anemones, and their allies are collectively named—in which the stinging-cells are of sufficient size and power to inflict severe pain on man himself. Aristotle was fully aware of this latter fact, when he named the Jelly-fishes and their allies *Acalephœ*, or 'Nettle-like' animals. And bathers and swimmers, through instinct, if not through zoological knowledge, generally and wisely contrive to give the Jelly-fishes a wide berth in their marine meanderings. The late Edward Forbes, in his humorsome manner, says of one species of jelly-fish, that, 'once tangled in its trailing "hair," the unfortunate who has recklessly ventured across the graceful monster's path, too soon writhes in prickly torture. Every struggle,' he continues, 'but binds the poisonous threads more firmly round his body, and then there is no escape;' for, as the naturalist informs us, even when the arms or tentacles are cast loose from the body of the jelly-fish, they 'sting as fiercely as if their original proprietor itself gave the word of attack.' The Abbé Dicquemare, an observant French naturalist, found that some species can sting only the more sensitive parts of the body, such as the eyes. But Forbes's

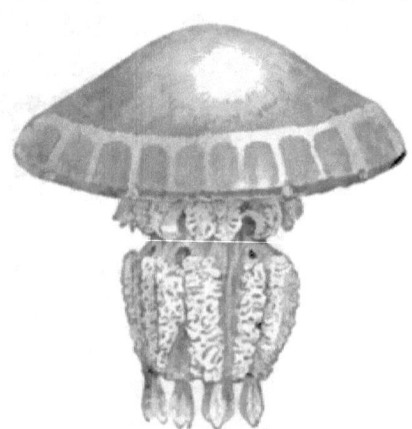

Fig. 71.—Rhizostoma.

remark on the Abbé's experiment, that most people would prefer 'keeping their eyes intact, to poking medusæ into them,' will coincide, we imagine, with the opinions of most of our readers. It is equally worthy of remark that 'appearances' in natural history, as in ordinary life, are apt to be 'deceptive;' and looking at the grace and beauty of the Jelly-fishes, one might hardly be disposed to credit them with such virulent powers.

The most notable offenders of the jelly-fish class, in respect of their stinging powers, are the *Physaliæ*, or Portuguese men-of-war, as they are popularly termed—a group of beautiful oceanic forms, met with floating far out at sea, especially in tropical latitudes, and presenting the appearance of a bladder-like structure, provided with a crest and trailing streamers, all coloured of the most ethereal and beautiful of hues. When the tentacles of a Physalia are allowed to come in contact with the human skin, the thread-cells—which are of large relative size—sting so severely, that the effects of the irritation may persist for a considerable time, and may give rise in some cases to very painful after-effects.

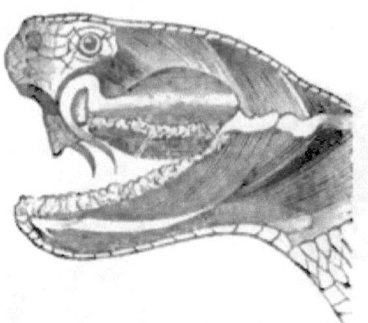

Fig. 72.—Head of Serpent.

The thread-cells in the tentacles of the common species of sea-anemones have no effect on the skin of man; but as the writer has frequently demonstrated, if the tentacle be allowed to touch the more delicate mucous membrane of the lips, a slight stinging sensation, accompanied by temporary numbness, may sometimes be felt.

Passing in review the higher groups of the animal kingdom, we find an endless variety of contrivances subserving offensive purposes, or limited to the milder purposes of defence. Shells, scales, and plates of every kind, with special modifications for special purposes, may thus readily be selected as examples; spines and allied armaments of all shapes and sizes; the poison-fangs and virulent secretions of centipedes and serpents, and the sting of spiders, scorpions, and bees, possessing sure and sometimes deadly effect on those they attack; and, in quadrupeds, strong claws and teeth united to equally powerful muscles—such are a few examples of the endless stores of weapons contained in animal armouries.

'FOOT-PRINTS ON THE SANDS OF TIME.'

LONGFELLOW'S well-known line which heads this chapter is by no means so metaphorical as is generally supposed. On the contrary, the geologist is able to point to numerous instances not merely of foot-prints which have been impressed on the sands of past ages, but to important discoveries regarding the past life of our globe, which have resulted from the study of these impressions. It may, therefore, form no uninteresting study if we endeavour, even in a cursory manner, to glance at some of the chief facts which the search amongst the fragments of other worlds has brought to light and disclosed.

About the year 1823 the Rev. Dr Duncan, a thoughtful and observant geologist, directed attention to the curious markings on certain sandstones in Dumfriesshire, which bore a singular resemblance to the foot-prints of reptiles of large size; and in 1834, foot-prints of very large size, and which exhibited a close likeness to the general shape of the human hand, were discovered in Saxony, in rocks of the same age as the Dumfriesshire sandstones. The resemblance of these foot-prints to the form of the human hand gave origin to the name of *Cheirotherium*, or 'hand-beast,' bestowed upon the then unknown animal, which had walked over the surface of the formation when it was soft and fresh. The discovery thus made, as we shall presently note, led to others of exceeding importance and interest; whilst the attention of geologists was in this manner directed to a new source of information regarding the animals of the past, the department which more especially investigated these impressions being named Ichnology, or 'foot-print lore.' It may appear a strange and wonderful thing that we are enabled from a mere foot-print, or a fragment of a bone or tooth, to build up and restore the entire frame of an animal. But in reality, the process of reconstruction is less wonderful than might be supposed, since the geologist merely takes into account two chief facts—which lie indeed at the root and foundation of his science—firstly, that the present is the key to the past; and secondly, that Nature is consistent and uniform in all her ways and works.

Bearing in mind these two chief facts, let us try to understand how they may be applied to unravel the mystery which overhangs the past of our earth, and to enable us to read the literal 'sermons in stones,' which are written so widely and unsparingly that even 'he who runs may read.' The foot-prints in the rock, to begin at the beginning, were evidently imprinted when the rock-surface was soft and fresh, and before it had become consolidated into the hard unyielding stone. How the impressions have kept their outline and form we shall presently note; but that we may form an idea of the circumstances under which foot-prints are impressed and formed, we may turn to the world at large, and survey the habits of various animals, and more especially of those which dwell near lakes or by the sea. If we walk along a sea-beach at the ebb of the tide, particularly where a layer of soft and fine mud has just been deposited at the farthest margin and slope of the shore, we may form some idea of the initial stages in the formation of foot-prints. Yonder, a sea-gull has just risen in flight, startled by our too near approach. If we carefully note the spot over which the gull has walked, we shall find abundant evidence of its presence in the numerous foot-prints it has left behind, clearly imprinted in the soft mud, which is bearing similar evidence in the shape of our own foot-prints to the presence of other and inquiring bipeds. When we scrutinise the impressions left by the gull, we may form some idea not only of the nature and kind of the bird itself, but also of its probable size. Thus we find the impression of three front toes, which we can also learn from the foot-print were united, as we know the front toes of the gull to be, by a web or membrane; whilst, from the faint impression of the hinder toe, we might hazard the guess that it was of small size, and unconnected with the other toes by the web. The manner in which the foot-prints succeed one another would afford us information regarding the two-legged condition of the creature, even supposing we did not possess exact knowledge of the existence of a lower race of animals which walked on two limbs; and, as will be presently exemplified by the consideration of the giant foot-prints of the Connecticut Valley in America, the inference as to the two-legged nature of the animal is by no means an unimportant observation.

Suppose, next, that a sandpiper or lapwing is seen running along the sea-beach, and that we examine the impressions left by this bird, we shall find the foot-prints in the latter case to differ materially from those of the gull. The imprints of the toes, like those of the gull, will be three in number, but no trace of a hinder toe will be seen; this latter member in the sandpiper being not only of small and insignificant size, but being raised off the ground on the back of the foot. And we may thus draw a number of reasonable inferences respecting the nature, structure, and habits of the birds which have left their impressions on the sea-beach, even although we may have no opportunity of observing the birds themselves. These examples, which might be extended to include other groups of animals, will suffice to shew the general principles on which the geologist investigates the foot-prints of the rocks. Settling, in the first instance, their animal nature, he proceeds to argue respecting the unknown from his knowledge of the known. The more perfectly he understands the present state of matters in the world, and as regards its living things, the more perfect and correct will be his deductions respecting the things and objects of the past. He uses his knowledge of the present, in other words, as the key which unlocks the puzzle of the past.

It may, however, be shewn that the principle thus illustrated applies equally well to the investigation of incidents in the past history of the earth other than those which belong to the higher animals. In the sand of the sea-shore we perceive countless worm-burrows, and the tracks of wandering crabs and other forms of marine life, plainly impressed on the yielding material. The observation of the wind-driven rain-shower which sends its drops deep into the soft mud, and of the wavy lines of sand-ripples—those

> Long waves on a sea-beach,
> Where the sand as silver shines—

which mark the gentle and declining action of the waves, may furnish us with the knowledge whereby we may account for markings on many of the rock-formations of our earth. For, when we disinter the masses of stone from a quarry, and see thereon represented the petrified sand-marks and wave-ripples, the marks of the wind-blown rain-drops, and the burrows of ancient worms, we can explain how these were formed from the knowledge of the existing sea-beach which the observant eye affords; and we may perchance be able in some cases to tell the direction whence that old rain-shower came, as the wind blew its drops against the then soft sand or clay. Again we are reading and explaining the past by the light of the present, and we have been tacitly assuming that nature has been uniform and consistent; since we have supposed that winds have blown, that rain has fallen in the past as in the present, and that the conditions and actions we see now, were those which existed and acted in the far-back and ancient past.

The manner in which these foot-prints and other marks are preserved, notwithstanding the pressure and consolidating process to which the soft materials are subjected in the course of rock-formation, is also fully explained to us by the simple and careful observation of nature. It was long ago pointed out that in many sea-beaches the composition of the sand and mud is admirably adapted for receiving and retaining impressions made upon it. The heat of the sun will further act in hardening the mud, and necessarily in fixing, as it were, the outlines of the foot-print. The returning waves, laden with soft particles of new matter, will fill up the impression as a mould is filled with the substance of which a cast is to be made; and each succeeding layer deposited above this first one, will tend simply to fix the mould and the cast firmly in place. Consolidation and pressure in due time succeed. The once soft material has, after the lapse of ages, become the solid rock; and when the block of sandstone from the quarry or the cliff is split asunder, we behold on the one face of our split portion the mould, and on the other face the material which has filled it up. Or, as may be seen on our coasts, the foot-prints and impressions may become filled up with fine sand which has been blown over the beach in clouds by high winds. This dry sand becomes moistened by the return of the tide, and new material is also added thereto, the cast being thus fixed gradually in the mould. Curious indeed it is to think that under our eyes are being thus formed the impressions which may afford matter for thought to future generations; whilst the subject no less impresses upon us the great importance in scientific research of noting what may appear at first sight to be very commonplace circumstances and very trivial things.

Perhaps the oldest foot-prints or traces of animal life that are known, exist in the Cambrian rocks. In these formations we find numerous worm-burrows and worm-tracks. Some of these burrows are straight (*Scolithus*); others are curved, such as those named *Histioderma*; and others again, such as *Arenicolites*, are looped, and open by two apertures on the surface. It is needless to remark that the animals which have constructed these burrows are entirely unknown to us, but the traces they have left behind present evidence of their existence and nature, as positive as that which would be afforded by the discovery of their remains. In these same Cambrian rocks occur impressions of peculiar kind, consisting in each case of a series of well-marked and somewhat rounded imprints, situated on each side of a middle furrow, exhibiting varied depths. These impressions exist in pairs, but there are no marks which would correspond to the imprint of toes or nails. During the period corresponding to the formation of the Cambrian rocks, the only actively organised animals were crustaceans of peculiar type and form, named *Trilobites*.

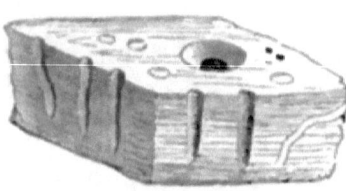

Fig. 73.—Worm Borings (Arenicolites): From Cambrian Rocks.

It is therefore a fair inference that these peculiar markings were produced by some large, extinct, crab-like creature possessing divided limbs and a prominent tail; the latter member forming the middle impression as its possessor slowly drew it along through the sand of that primitive shore.

The more definite and more readily determined fossil foot-prints of higher animals are exemplified very clearly by the case, already mentioned, of the *Cheirotherium*—the unknown animal whose foot-prints were discovered in 1834 near the village of Hessburg in Saxony. These foot-prints bear, as previously mentioned, a close

Fig. 74.—Labyrinthodon Foot-prints.

resemblance to the general form of the human hand. They were also found to occur in sandstones in Warwickshire and Cheshire in England; and were discovered on the slabs of stone being split asunder, when the one face of the stone exhibited the impression, the other face shewing the foot-prints in relief. The impressions of the hinder feet each measure eight inches in length, and five inches in breadth; and close by each of these larger imprints, and at a regular distance of an inch and a half before the latter, a smaller foot-print, corresponding to the impression of the fore-foot, exists. The distance between one pair of foot-prints and the succeeding pair is about fourteen inches; and each foot-print shews the small thumb or outer toe to have been borne on the outermost side of each foot.

A closer examination of the foot-prints revealed the fact that they most closely resembled those of the well-known salamanders, since the short outer toe of each hind foot projected almost at a right angle to the middle toe; whilst it was also determined that no known reptile or amphibian animal could form an impression corresponding in all points to that of the Cheirotherium. As time passed, however, and as further discoveries in the extinct life of the rocks were made, the fossil remains of animals of large size were found, the animals to which these remains belonged being named *Labyrinthodons*, from the complicated nature of the teeth. The skull of one species of these animals measures three feet in length and two feet in breadth, and indicates the development of an animal, the dimensions of whose body must have been sufficiently large to have produced the huge foot-prints of Cheirotherium. The body in these creatures was covered with bony plates, whilst the limbs were weak in proportion to the size of the body, and a long lizard-like tail existed. In the Labyrinthodons, then, the remains of which occur in the rocks in which the Cheirotherium footprints were found, we find the authors of the latter impressions; the inferences drawn from the foot-prints as to the peculiar nature of the animals which formed them, being fully borne out by the discovery of the animals themselves.

Passing by the foot-prints found in the New Red Sandstone rocks of Dumfriesshire, and which are supposed to be those of extinct tortoises, we arrive at the famous impressions, supposed to be those of the feet of birds, which were first noticed in 1835 in the Triassic sandstones of the Connecticut Valley in the United States.

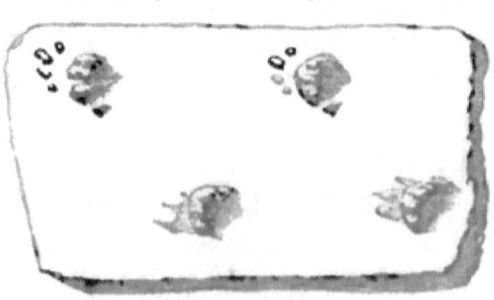

Fig. 75.—Foot-prints of a Tortoise: From the Permian Sandstone of Annandale.

These foot-prints were discovered by Dr Deane of Greenfield, U.S., and were by him described as those of birds. This announcement elicited considerable surprise and evoked much interest, for two reasons. Thus, firstly, the discovery of these impressions had brought to light the first evidences of the existence of birds on the earth; and secondly, the large size of the foot-prints shewed

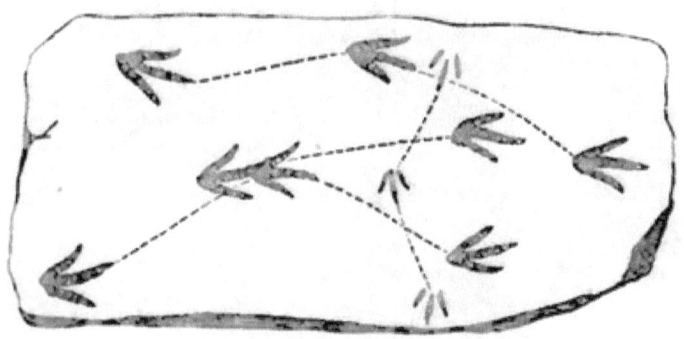

Fig. 76.—Foot-prints of Birds:
On the Oolitic Sandstones of Connecticut.

them to have belonged to birds, compared with which, the ostrich, largest of living birds, might be regarded as a mere pigmy. The largest of these impressions measures twenty-two inches in length, and Dr Hitchcock, who originally described the impressions, submitted his opinion that during the period in which the sandstones were formed and deposited, birds at least four times larger than the ostrich must have existed. The foot-prints succeeded each other at definite intervals, and differed from each other simply with the amount of difference which exists between a right and left foot. Each impression exhibited the imprints of three toes, which, like those of birds, diverge or spread widely outwards. A space measuring twelve inches intervened between the tips of the inside and outside toe-marks; whilst the impressions of claws or nails were visible at the extremities of the foot-print. As in birds, the mark of the 'foot' proper was represented by a kind of double impression, produced by the end of the leg. The number of the joints or separate bones in the inner toe was seen to be three; four joints existed in the middle, and five in the outer toe; the joints in these huge feet exactly corresponding in number to those found in the toes of existing birds. So clearly were

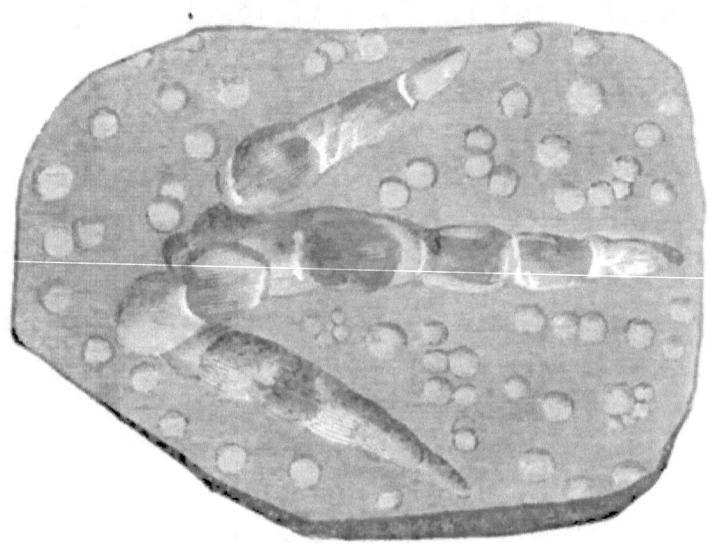

Fig. 77.—Foot-prints of Bird, and Rain-drops.

the various parts of the foot represented in the impression, that even the roughened surface of the skin which covered the under surface of the toes was duly reproduced in the foot-print.

Thus far, the evidence derived from the investigation of living birds would seem to bear out the inference that the foot-prints of the Connecticut Valley were those of these bipeds. But the consideration of the instance before us serves to shew that the evidence must in all cases be weighed with great care, and in view of all the circumstances which the higher knowledge of succeeding years may bring. We have thus noted the evidence of the bird-character of the foot-prints to consist in their two-footed character; in their exhibiting the same number of toes as very many birds exhibit; and in the toes each possessing the same number of joints as is represented in those of living birds. The evidence is thus apparently complete enough as it stands; and in view of the fact that no known animal save a bird could produce impressions of corresponding nature.

In these same sandstones, however, four-toed foot-prints have since been met with, and the fuller investigation of some extinct forms of reptiles has strongly suggested the probability that the three-toed impressions may in reality be of reptilian nature. The further knowledge that many of these extinct reptiles undoubtedly possessed the power of walking on two limbs, strengthens this supposition. Thus there is one known extinct reptile named *Compsognathus*, which possesses hind-limbs of a type remarkably like those of the ostriches. And another well-known extinct reptile, *Iguanodon* by name, which must have attained a length of from fifty to sixty feet, in all probability possessed a power of walking on two limbs; the hinder limbs being exceedingly large and powerful, as compared with the front members. Iguanodon had three toes on the hinder feet, and considering the close similarity between the bird and reptile type—dissimilar as these animals appear to be—it is not surprising to find many geologists leaning to the supposition that the Connecticut foot-prints may, after all, be those of reptiles, or indeed of forms which may be regarded as somewhat intermediate between the bird and reptile classes.

Having glanced at some of the chief facts regarding fossil foot-prints, we may next turn to the consideration of the allied subject of fossils themselves. The study of foot-prints links itself, as we have seen, in the most complete fashion with that of the animals, living or extinct, which may be regarded as having produced the impressions. The name 'fossil,' as every one knows, is given to the petrified remains of animals and plants preserved in the rock-formations of our earth, and which are composed, for the most part, of the same substance as that of the rock in which they are imbedded. The older geologists, not realising that the fossils they met with in the rocks represented simply the strays and fragments of the life which inhabited our world in past ages and former epochs, regarded these relics with somewhat of superstitious awe, and still later speculated very loosely, and in some cases absurdly, regarding their origin. Fossils were thus believed by some to be produced by the influence of the heavenly bodies, by others to be 'freaks of nature;' whilst, when their true nature had been determined, the entire collection of fossils were believed to have been deposited by and formed at the time of the Deluge. The careful study of fossils has, however, served to clear away most of the errors which the older geologists propagated, and has besides afforded us much valuable information regarding the past history of the world. We are enabled, in fact, from a study of fossils, to group the rocks into very definite periods, each period or epoch succeeding another period, and each possessing its due and stated place in the entire series of rocks. Whilst we thus also arrive at some intelligent conceptions of the age of rocks, and of the length of time which has elapsed since the formation of any particular group of formations. When the geologist, for example, finds shells of fresh-water kind in rocks, he assumes that these rocks must have been formed in lakes, or from the sediment of fresh waters. The discovery of fossil corals and fishes, and of other animals which are known to inhabit the sea, would shew that the rocks in which such fossils were found, represented a marine or salt-water formation; and the geologist is also able, from an inspection of the fossils of any series of rock-beds, to say if the rocks have been formed in brackish situations, exemplified by the mouth or estuary of a large river, where the marine and fresh-water deposits merge into one another.

It is, however, less with the evidences derived from fossils, than with the manner of their formation, that we purpose now to deal. The natural conclusion to which the observer must come when he sees the imprint of a shell in a rock, or when he beholds the petrified shell itself, is that the shell once formed part of a living animal which inhabited it, and that in some way or other it became enclosed in the rock-materials when these latter were soft, and before they had become consolidated. The shell has in fact become rock, along with the rock, and is now to be regarded as an essential part of the formation. Similarly, when we see a petrified bone or a tooth, we assume that the one once formed part of an animal's skeleton, and that the other once grew in an animal's mouth. If we meet with a fossil tree, or with the impress of some delicate fern—'nature-printed' as it were on the rock-surface—we assume that it once grew and waved its fronds in the air and sunlight, as do its living neighbours of our own world and day. It is necessary to mention these facts and inferences, because some persons have now and then suggested that the rocks might have been formed with the fossils in them. This supposition, it need hardly be said, is one entirely unsupported by common sense, or by the slightest particle of scientific evidence.

The simplest and most complete manner in which a fossil may be formed is that where the actual body of the plant or animal is replaced by mineral matter. This process gives us the most perfect fossils, and its results are well seen in the complete preservation of the wood of trees, which, when minutely examined, is seen to present every part of its structure perfectly reproduced in its fossil state. The rings seen in the trunk of a tree, the fibrous structure of the wood, and even the delicate markings on the minute vessels of the plant, may be seen perfectly preserved in fossil plants. By grinding down thin sections of fossil plants, and by submitting them to microscopic examination, the botanist has often been enabled to determine the particular groups to which the plants have belonged. The bodies of animals, and hard parts of animals, have also been preserved in this complete manner, enabling the geologist to determine with the greatest accuracy the features of some extinct being, utterly unknown in the existing world of life. This first process begins with the animal or plant falling into water, and being conveyed sooner or later to the sediment which is being formed at the bottom of a lake, or in the bed of the sea. The water-formed or aqueous rocks, or those which have been formed from the sediment of rivers, lakes, and seas, are those which alone contain fossils; the fire-formed or igneous rocks—such as are represented by the lava and other substances ejected from volcanoes—as might be expected, containing no traces of life. Being entombed amongst the soft sediment, the body of the animal or plant begins to decay and to fall to pieces; this process, the inevitable result of death, proceeding somewhat more slowly, however, than under ordinary circumstances. As decay proceeds, however, and as particle after particle of the once-living matter disappears, these atoms are in the same gradual manner replaced by particles of the mineral matter by which it is surrounded. It is, as has well been remarked, as if a house were being rebuilt 'brick by brick, or stone by stone, a brick or a stone of a different kind having been substituted for each of the former ones, while the shape and size of the house, the form and arrangement of its rooms, passages, and closets, and even the number and shape of the bricks and stones, remained unaltered.' The soft sediment itself next undergoes alteration and changes of great extent. The work of consolidation proceeds apace, and in the general chaos and confusion of the past, the rock-beds may have been far removed from their original site. And when, latest of all, man, with the inquiring hammer of the geologist, or the sturdy pickaxe of the quarryman, splits asunder the rock-layers, he meets with the preserved relic of a far-back past—the memento, it may be, of a time when he himself was not, and when the present order of the world was yet in the shadowy and distant future.

A less complete manner of preservation is that in which internal casts of living beings or their parts are found. Thus a shell, such as that of a cockle, may be supposed to fall into the soft deposit, which fills up the interior; whilst around the shell itself the same matter will also fit closely, so as literally to bury the object. When such a fossil as this is examined, we should find the shell to be completely solid. A cast of the interior of the shell would clearly be formed by the matter which had gained admittance within the shell; whilst the matter amidst which the shell was buried, would present us with a cast of the outside surface of the shell. Very frequently it happens that the limy substance of the shell itself is dissolved away by chemical action soon after its entombment, and we are thus left with the casts just alluded to, as the only remaining traces of the animal. Lastly, we may meet with mere impressions of the bodies of animals, just as we have noted the imprints of their footsteps to occur, and occasionally we may gain information respecting animals and plants, which, but for the fact of their mere impressions having been found, would have been wholly unknown to us. There are many delicate impressions of extinct sea-weeds, zoophytes, and delicate ferns, affording us copious stores of information regarding the once-living organisms, which must have been so delicate that their preservation in any other fashion must be regarded as an impossibility. It is, as could readily be supposed, only the *hard* parts of animals and plants which have, as a rule, been preserved to us as fossils; the soft parts—save for their being preserved as impressions—and save under exceptional circumstances, disappearing under the influence of decay and pressure. Noting these facts, it is interesting and curious to note that we possess evidence of the existence, in the far-back past, of those softest of animals the jelly-fishes, in the shape of more than one delicate impression left on the soft mud of some old sea-beach, on which the jelly-fish was tossed, as are its modern neighbours, to perish and die. In the fine Lithographic Slates of Solenhofen, in Bavaria, impressions of jelly-fishes have thus been observed.

A general survey of the life of the past shews us that living beings have been developed in a gradual manner. The oldest rocks contain the simplest of organisms, and it is only as we advance upwards and approach nearer to our own age that we find the higher animals represented. Very curious and weird-like are many of the forms which the geologist has disinterred from the safe keeping of Mother Earth; and as he pieces together the fragments of the puzzle of life, he can see enough, notwithstanding the imperfection of his knowledge, to form some idea of the wondrous concourse and succession of living beings which have peopled this world—some species persisting to our own day, whilst others have appeared only, as it were, to disappear. Particularly is this the case with many extinct and gigantic reptiles, which appear to have had, speaking geologically, a very brief existence, and to have presented some very strange anomalies in structure when compared with their nearest living representatives. Thus, in the rocks known to geologists as the Trias, the Oolite, and the Chalk formations, the remains of huge reptiles abound. Compared with the size of their extinct ancestors, the largest of our modern reptiles appear small and insignificant. There, for example, is the *Ichthyosaurus*, or great 'fish-lizard,' which attained a length of at least twenty or thirty feet, and which, as revealed by the structure of its spine, must have been able to swim as actively as any fish. This huge

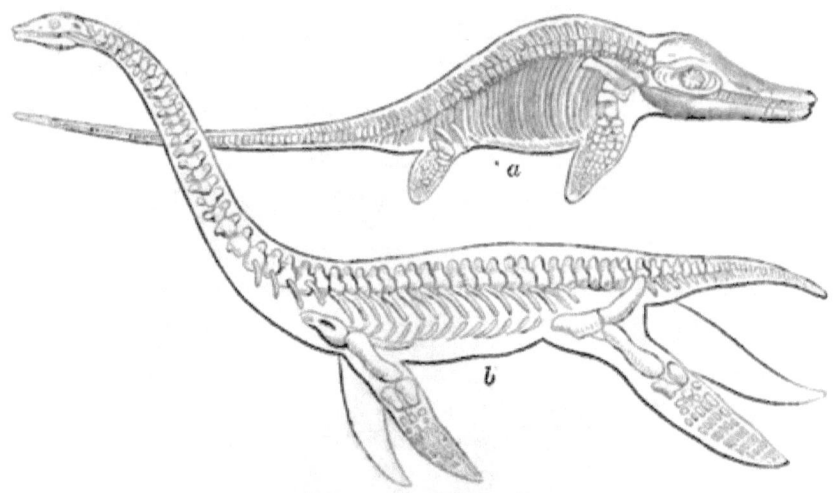

Fig. 78.—Fossil Reptiles:
a, Ichthyosaurus; *b*, Plesiosaurus.

reptile probably ventured far out to sea, swimming by aid of its paddle-like limbs, and the tail-fin with which in all probability it was furnished. The jaws were plentifully provided with teeth, and the eye was supported by a ring of bony plates in front; this structure enabling the animal 'to descry its prey,' as Dr Buckland remarks, 'at great or little distances, and in the obscurity of night, and in the depths of the sea.' Or look at its neighbour the *Plesiosaurus*, with which doubtless the 'fish-lizard' had many a combat, which attained a length of from twenty to thirty feet, and which 'to the head of a lizard' unites 'the teeth of a crocodile, a neck of enormous length resembling the body of a serpent, a trunk and tail having the proportions of an ordinary quadruped, the ribs of a chameleon, and the paddles of a whale.' This animal must also have been an active free-swimming creature, pursuing its prey, consisting chiefly of fishes, through the waters of the primitive oceans in which it lived; whilst, when it lurked by the shores of the sea, or at the estuaries of rivers, its long neck would enable it to descry its prey as it peered above the tall reeds and water-plants, and would also serve as an effective instrument to secure the finny prey. No less curious than the preceding forms is the *Pterodactyl*, or 'wing-finger,' a reptile which

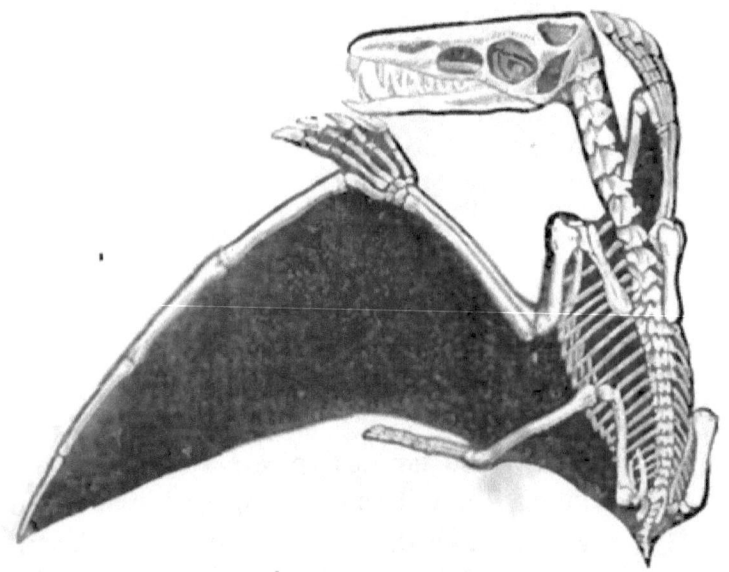

Fig. 79.—Fossil Skeleton of Pterodactyl:
The black portions represent the extent of the wing-membrane.

possessed a power of flight equal to that exhibited by our bats, and which in many cases must have attained an expanse of wing measuring twenty feet. The jaws were provided with teeth, and the eyes must have been large and prominent. The wing-membrane was formed as in bats by an expansion of the skin of the body, and was supported chiefly by an enormously long outer finger; the other fingers being shut. This membrane was attached to the sides of the body, and to the hind-limbs, and extended also between the hind-limbs and tail. The latter members were very small and of weak nature, so that on the ground these reptiles must have been very helpless animals, and must have crawled along slowly, and after the awkward fashion of existing bats. But in the air, on the other hand, they must have been both fleet and active; and it is curious to note that the only living reptile which possesses a flying-membrane is the little lizard, named the 'flying dragon' of the East Indies; and even this animal possesses no power of true flight, but simply takes flying leaps from tree to tree; the expanded skin of the sides of the body acting simply as a kind of parachute, and supporting it temporarily in the air.

There again is the giant *Cetiosaurus*, or 'whale-lizard,' which fairly ranks first amongst these old reptilian giants, measuring, as it does, from sixty to seventy feet in length, whilst some of the joints of its tail measure seven and a half inches across by five and a half inches in length. This huge monster must apparently have lived a life partly on land and partly in water, and probably fed on the vegetation which bordered its aquatic haunts. Whilst, if we leave the reptiles and pass to the domain of the birds, we shall find that there were also bird-giants in these days. The bones found in the Recent or newest deposits of New Zealand shew us that a race of gigantic wingless birds inhabited that land, and in all probability within the human period, as suggested by the Maori traditions and legends. One species of these giant-birds, known as the Dinornis, stood, for example, upwards of ten feet in height, judging from the height of the shin-bone, which measures upwards of a yard in length; and the bones of the toes of another species, to quote Professor Owen's words, 'almost rival those of the elephant.' In Madagascar are found the remains of the *Æpiornis*, a bird as large as the extinct New Zealand forms. Some of the eggs of the Æpiornis have also been met with in a fossil state. These eggs are worthy of the giants which produced them; since each measures from thirteen to fourteen inches in long diameter, and is estimated to be as large as six ostrich eggs, or one hundred and forty-eight hen's eggs.

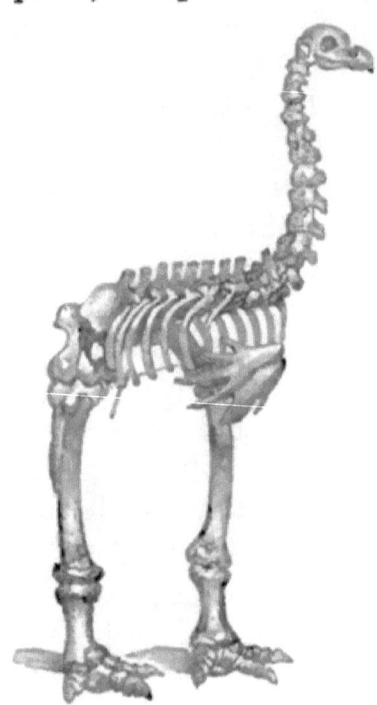

Fig. 80.—Skeleton of Dinornis.

Even the quadrupeds of the past would appear to have outrivalled their modern neighbours in size. What would be thought, for example, of a modern kangaroo the head of which measured three feet in length? Yet such are the dimensions of the extinct *Diprotodon* of the Australian bone-caves. The little armadillos of the South American forests were represented in like forests of the past by the great *Glyptodon*, a giant armadillo, which attained a length of nine feet. And the sloths of these same forests find their representatives of the past in the shape of the *Mylodon* and *Megatherium;* the former attaining a length of eleven, and the latter of eighteen feet. In addition to these wonders as regards size, the study of the past also reveals to us curiosities in the way of changes in the distribution of animals. The elephants of the past were not confined to Asia and Africa, as are their existing representatives, but roamed over the whole world, and were even found in the shape of the great hairy mammoths—the remains of which are still met with packed and preserved in the Siberian ice—amidst the cold and snow of the northern regions. The 'British Lion' is by no means a metaphorical creature in the eye of a geologist, since he can point to a comparatively recent period in the geological history of Britain, when our caves were tenanted by a 'Cave Lion,' exactly resembling the existing 'king of beasts.' And no less surprising is it to find that hyænas

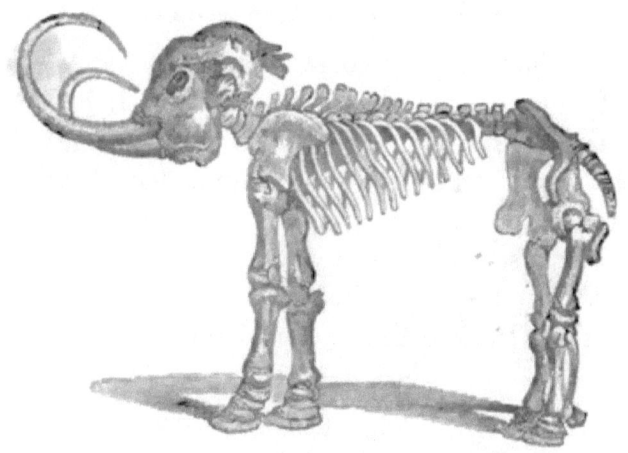

Fig. 81.—Skeleton of Mammoth.

and bears, the mammoth, elephant, and even the rhinoceroses, were represented in the British woods and forests of the past, and were, more curious still, beheld by the primitive men who first peopled Western Europe—a fact ascertained by the discovery of fragments of horns and tusks on which rude drawings of reindeer and mammoths were scratched by the early inhabitants of the Continent.

No more fascinating study can well be presented to the

human mind than that which investigates the nature and conditions of past life, and which seeks to read in the 'record of the rocks,' the biography of the world in which we live. The truly wondrous nature of these changes can be fully perceived only by the patient observation of nature; and vividly indeed does the geologist realise the beauty and truth implied in the Laureate's words:

> There rolls the deep where grew the tree.
> O earth, what changes hast thou seen!
> There where the long street roars, hath been
> The stillness of the central sea.
>
> The hills are shadows, and they flow
> From form to form, and nothing stands;
> They melt like mist, the solid lands,
> Like clouds they shape themselves and go.

And if one thought may come more prominently forward than another amidst studies of this nature, it is that which suggests the idea, that beneath all the changes through which the earth and its living things have passed, there may be traced not only a singular uniformity of plan and purpose, but also the wisest adaptation to the wants and ways of the creatures with which a wise and beneficent Mind has from time to time peopled the universe.

THE END.

Edinburgh: Printed by W. & R. Chambers.

www.ingramcontent.com/pod-product-compliance
Lightning Source LLC
Chambersburg PA
CBHW020904230426
43666CB00008B/1308